AF352186

AMERICAN MATHEMATICAL SOCIETY TRANSLATIONS

Series 2

Volume 133

Thirteen Papers in Analysis

by

R. R. Suncheleev

F. A. Shamoyan

V. È. Katsnel'son

A. È. Eremenko

L. A. Markushevich

N. B. Tikhomirov

S. Baron

E. G. Antsiferov

Yu. P. Rybakov

A. T. Fomenko

A. B. Katok

V. I. Bilenko

M. Z. Berkolaĭko

AMERICAN MATHEMATICAL SOCIETY
PROVIDENCE, RHODE ISLAND

Translation edited by BEN SILVER

1980 *Mathematics Subject Classification* (1985 *Revision*). Primary 22E35, 26A16, 30D55, 31A05, 34A20, 34C35, 34D20, 35JXX, 41AXX, 42A50, 42B15, 45G10, 46E35, 47B35, 54H20, 58FXX, 65D05, 65K05, 81E08; Secondary 17B99, 30D35, 30G30, 45-04, 49D15, 65R20, 65V05.

Library of Congress Cataloging-in-Publication Data
Thirteen papers in analysis.

(American Mathematical Society translations; ser. 2, v. 133)
1. Numerical analysis. I. Suncheleev, R. R. II. Series.
QA3.A572   ser. 2, Vol. 133   510 s   [511]   86-26455
[QA297]
ISBN 0-8218-3109-7

COPYING AND REPRINTING. Individual readers of this publication, and nonprofit libraries acting for them, are permitted to make fair use of the material, such as to copy an article for use in teaching or research. Permission is granted to quote brief passages from this publication in reviews, provided the customary acknowledgment of the source is given.

Republication, systematic copying, or multiple reproduction of any material in this publication (including abstracts) is permitted only under license from the American Mathematical Society. Requests for such permission should be addressed to the Executive Director, American Mathematical Society, P.O. Box 6248, Providence, Rhode Island 02940.

The appearance of the code on the first page of an article in this book indicates the copyright owner's consent for copying beyond that permitted by Sections 107 or 108 of the U.S. Copyright Law, provided that the fee of $1.00 plus $.25 per page for each copy be paid directly to the Copyright Clearance Center, Inc., 21 Congress Street, Salem, Massachusetts 01970. This consent does not extend to other kinds of copying, such as copying for general distribution, for advertising or promotional purposes, for creating new collective works, or for resale.

Copyright ©1986 by the American Mathematical Society. All rights reserved.
Printed in the United States of America.
The American Mathematical Society retains all rights
except those granted to the United States Government.
The paper used in this book is acid-free and falls within the guidelines
established to ensure permanence and durability. ∞

Contents

Russian Contents[*]

---

[*]The American Mathematical Society scheme for transliteration of Cyrillic may be found at the end of index issues of *Mathematical Reviews*.

Amer. Math. Soc. Transl.
(2) Vol. **133**, 1986

Harmonic Analysis on p-adic Lie Groups

UDC 519.53

R. R. SUNCHELEEV

Let $\mathbf{Z}_p$, $\mathbf{Q}_p$, and $\mathbf{C}_p$ denote the ring of p-adic integers, the field of p-adic numbers, and the completion of the algebraic closure of $\mathbf{Q}_p$, respectively. As in [1], π_p denotes the group dual to $\mathbf{Z}_p$. It is known that on p-adic Lie groups there does not exist an invariant bounded measure with p-adic values. However, it is shown here that on the groups $\mathbf{Z}_p$ and $\mathrm{Aff}(\mathbf{Z}_p)$ there exist spaces of functions admitting invariant integration. Analogues of group algebras are constructed for these groups, and their structure is studied.

§I. The group $\mathbf{Z}_p$

Denote by $\mathrm{Loc}(\mathbf{Z}_p)$ the space of continuous locally constant functions on $\mathbf{Z}_p$ with values in $\mathbf{C}_p$. On $\mathrm{Loc}(\mathbf{Z}_p)$ we define a functional I_0 by

$$I_0(f) = \lim_{n \to \infty} \frac{1}{p^n} \sum_{i=0}^{p^n-1} f(x_i), \qquad x_i \in i + p^n \mathbf{Z}_p.$$

This limit obviously exists and does not depend on the choice of the x_i. In what follows we often write $\int_{\mathbf{Z}_p} f(x)\,dx$ instead of $I_0(f)$.

Let π_p be the group dual to $\mathbf{Z}_p$ in the sense of Pontryagin duality. It can be imbedded in $\mathbf{C}_p$:

$$\pi_p = \{\omega \in \mathbf{C}_p \colon \omega^{p^n} = 1,\ n = 0, 1, 2, \ldots, \}.$$

We define the Fourier transformation F on $\mathrm{Loc}(\mathbf{Z}_p)$:

$$[Ff](\omega) = \int_{\mathbf{Z}_p} f(x)\omega^{-x}\,dx.$$

1980 *Mathematics Subject Classification* (1985 *Revision*). Primary 22E35.
Translation of Dokl. Akad. Nauk UzSSR 1981, no. 11, 9–11; MR **83k**:22027.

©1986 American Mathematical Society
0065-9290/86 $1.00 + $.25 per page

THEOREM 1. *If $f \in \mathrm{Loc}(\mathbf{Z}_p)$, then*

$$f(x) = \sum_{\omega \in \pi_p} [Ff](\omega)\omega^x,$$

where only finitely many terms are nonzero in this sum.

Define the norm $\|\cdot\|_L$:

$$\|f\|_L = \max_{\omega \in \pi_p} \left| \int_{\mathbf{Z}_p} f(x)\omega^{-x}\, dx \right|,$$

on $\mathrm{Loc}(\mathbf{Z}_p)$ and denote by $CL(\mathbf{Z}_p)$ the completion of $\mathrm{Loc}(\mathbf{Z}_p)$ in this norm. Obviously, $CL(\mathbf{Z}_p)$ consists of all functions of the form $f(x) = \sum_{\omega \in \pi_p} \lambda_\omega \omega^x$, where $\lim_{\nu(\omega) \to \infty} \lambda_\omega = 0$. It is clear that $\|I_0\| = 1$ and the functional I_0 extends to a continuous functional I on $CL(\mathbf{Z}_p)$ that is translation invariant.

We define on $CL(\mathbf{Z}_p)$ the convolution operation

$$[f * g](x) = \int_{\mathbf{Z}_p} f(y)g(x - y)\, dy.$$

It is not hard to see that $CL(\mathbf{Z}_p)$ is a Banach algebra with convolution as multiplication, the usual addition, and multiplication by scalars.

Denote by $B(\pi_p)$ the space of functions on π_p with values in $\mathbf{C}_p$ that tend to zero at infinity. We provide $B(\pi_p)$ with the norm $\|\varphi\|_B = \max_{\omega \in \pi_p} |\varphi(\omega)|$. Then $B(\pi_p)$ is a Banach algebra with the usual operations of addition and multiplication and with the specified norm.

The Fourier transformation on $\mathrm{Loc}(\mathbf{Z}_p)$ extends naturally to the Fourier transformation $F: CL(\mathbf{Z}_p) \to B(\pi_p)$.

THEOREM 2. *The Banach algebras $CL(\mathbf{Z}_p)$ and $B(\pi_p)$ are isometrically isomorphic.*

§II. The group $\mathrm{Aff}(\mathbf{Z}_p)$

Let $G = \mathrm{Aff}(\mathbf{Z}_p)$ be the group of affine transformations of $\mathbf{Z}_p$. It can be identified with the group of matrices of the form $\left(\begin{smallmatrix} a & b \\ 0 & 1 \end{smallmatrix}\right)$, where $a \in \mathbf{Z}_p^x$ and $b \in \mathbf{Z}_p$ [2]. The subgroups

$$G_n = \left\{ \begin{pmatrix} a & b \\ 0 & 1 \end{pmatrix},\ a \equiv 1 \ (\mathrm{mod}\, p^n),\ b \equiv 0 \ (\mathrm{mod}\, p^n) \right\}_{n=1,2,\ldots}$$

are normal subgroups of G. On the algebra of sets generated by all possible cosets of G by G_n we define a measure with values in $\mathbf{C}_p$:

$$\mu(gG_n) = \frac{1}{p^n(p^n - p^{n-1})}.$$

Obviously, μ is invariant and unbounded, and $\mu(G) = 1$.

Let $E_k = \{\sum_i \lambda_i \omega_{ki}^x, \ \lambda_i \in \mathbf{C}_p, \ \omega_{ki}$ a primitive p^kth root of unity$\}$. An irreducible representation T_k of the group G acts in E_k according to the formula $[T_k(a,b)f](x) = f(ax+b)$ [3]. Define a norm on E_k by

$$\left\| \sum \lambda_i \omega_{ki}^x \right\| = \max |\lambda_i|.$$

Then it is clear that $\|T_k(g)\| = 1$, $g \in G$.

Denote by $\mathrm{Loc}(G, \mathbf{C}_p)$ the space of locally constant functions on G (that is, functions f for which there exists a normal subgroup $G_n \subset G$ such that f is constant on the cosets of G by G_n). If $f \in \mathrm{Loc}(G, \mathbf{C}_p)$, then $\int_G f(g)\,d\mu(g)$ is defined in the natural way.

We define the Fourier transformation F on $\mathrm{Loc}(G, \mathbf{C}_p)$:

$$[Ff](k) = \int_G f(g)T_k(g^{-1})\,d\mu(g), \qquad [Ff](k) \in L(E_k),$$

where $L(E_k)$ is the space of linear operators on E_k.

THEOREM 3. *The equality*

$$f(g) = \sum_k (p^k - p^{k-1})\,\mathrm{tr}\left[\int_G f(g_1)T_k(g_1^{-1})\,d\mu(g_1) \circ T_k(g)\right].$$

holds for each function $f \in \mathrm{Loc}(G, \mathbf{C}_p)$.

Let $CL(G)$ be the completion of $\mathrm{Loc}(G, \mathbf{C}_p)$ in this norm. The functional $f \to \int_G f(g)\,d\mu(g)$ and the Fourier transformation F extend naturally to $CL(G)$. Define on $CL(G)$ the convolution operation

$$[f * \varphi](g) = \int_G f(gg_1^{-1})\Phi(g_1)\,d\mu(g_1).$$

It is not hard to see that $(CL(G), +, *)$ is a Banach algebra.

Let $\mathcal{B}$ be the subalgebra of $\prod_k L(E_k)$ consisting of all sequences that tend to zero in the norm; $\mathcal{B}$ becomes a Banach algebra with the norm $\|\{A_k\}\|_{\mathcal{B}} = \max_k \|A_k\|$.

THEOREM 4. *The Banach algebras $CL(G)$ and $\mathcal{B}$ are isometrically isomorphic. If $f \in CL(G)$, then*

$$f(g) = \sum_k (p^k - p^{k-1})\,\mathrm{tr}\left[\int_G f(g_1)T_k(g_1^{-1})\,d\mu(g_1) \circ T_k(g)\right].$$

BIBLIOGRAPHY

1. C. F. Woodcock, *Fourier analysis for p-adic Lipschitz functions*, J. London Math. Soc. (2) **7** (1974), 681–693.

2. A. A. Kirillov and R. R. Suncheleev, *Algebra of measures on the group of affine transformations of the p-adic interval*, Dokl. Akad. Nauk UzSSR **1975**, no. 2, 3–4. (Russian)

3. R. R. Suncheleev, Author's Summary of Candidate's Dissertation, Tashkent, 1976. (Russian)

Translated by H. H. MCFADEN

Amer. Math. Soc. Transl.
(2) Vol. **133**, 1986

Toeplitz Operators and Division by an Inner Function in Some Spaces of Analytic Functions

UDC 517.73

F. A. SHAMOYAN

§**1.** Let $\mathbf{D}$ be the unit disk on the complex plane, Γ its boundary, and H^p the Hardy class with the usual L^p-norm. If $f \in H^p$, then f admits the canonical factorization

$$f(z) = cz^n \prod_{k=1}^{+\infty} \frac{z_k - z}{1 - \overline{z}_k z} \frac{\overline{z}_k}{|z_k|} \exp\left\{ -\int_{-\pi}^{\pi} \frac{e^{i\theta} + z}{e^{i\theta} - z} d\mu(\theta) \right\}$$

$$\cdot \exp\left\{ \frac{1}{2\pi} \int_{-\pi}^{\pi} \frac{e^{i\theta} + z}{e^{i\theta} - z} \ln |f(e^{i\theta})| \, d\theta \right\}$$

$$= B_f(z) S_j(z) Q_j(z), \qquad z \in \mathbf{D},$$

where $\{z_k\}_1^{\infty}$ is the set of zeros of f in $\mathbf{D}$ and $d\mu$ is a nonnegative singular measure. Following Beurling (cf. [1]), we shall call $J_f = B_f S_f$ the inner part and Q_f the outer part of f. An inner function $\tilde{J} = \tilde{B}\tilde{S}$ is called a divisor of $f \in H^p$ if $B_f S_f \tilde{J}^{-1}$ is inner.

Let X be a space of analytic functions, $X \subset H^1$. In several problems of analysis an important role is played by assertions stating that if $f \in X$ and $\tilde{J}$ is a divisor of f, then $f\tilde{J}^{-1}$ belongs to X (cf., for example, [2]). For a detailed exposition of the state of the art of this question, see [2]–[4]. In [3] and [4] it has been established, for some spaces of analytic functions, that for any $h \in H^{\infty}$ the function

$$g(z) = P(\overline{h}f)(z) = \frac{1}{2\pi i} \int_{\Gamma} \frac{f(\varsigma)\overline{h}(\varsigma)}{\varsigma - z} d\varsigma, \qquad z \in \mathbf{D},$$

is not less smooth than f. We note that the study of the indicated Toeplitz operators in spaces of functions analytic in the disk and smooth up to the boundary has other interesting applications as well (cf. [2]). The goal of this paper is to

1980 *Mathematics Subject Classification* (1985 *Revision*). Primary 30D55, 47B35.
Translation of Akad. Nauk Armyan. SSR Dokl. **76** (1983), 109–113; MR **84j**:30093.

©1986 American Mathematical Society
0065-9290/86 $1.00 + $.25 per page

show that the operators $T_h(f) = P(\overline{h}, f)$ are bounded in the following spaces of analytic functions.

a) *Spaces of functions with derivatives in the classes of Dzhrbashyan.* Following Dzhrbashyan [5], [6], we denote by $H^p(\alpha)$, $-1 < \alpha < +\infty$, $0 < p < +\infty$, the class of functions f holomorphic in the disk for which

$$\|f\|_{H^p(\alpha)} = \left(\frac{\alpha+1}{\pi} \int_{\mathbf{D}} (1 - |\varsigma|^2)^\alpha |f(\varsigma)|^p \, d\sigma_2(\varsigma) \right)^{1/p} < +\infty.$$

A systematic study of these spaces was begun in [5] and [6]. Moreover, for a natural number n, the symbol $H^p_n(\alpha)$ will denote the class of functions f whose nth derivative belongs to $H^p(\alpha)$. It is easy to see that for $1 \le p < +\infty$ the norm

$$\|f\|_{H^p_n(\alpha)} = \max_{0 \le k \le n} \|f^{(k)}\|_{H^p(\alpha)}$$

turns $H^p_n(\alpha)$ into a Banach space. For $n > (\alpha + 1)/p$ we have the imbedding $H^p_n(\alpha) \subset H^1$ with a continuous imbedding operator.

b) *Holomorphic functions with boundary values from the Besov classes.* Let $1 \le p, q < +\infty$ and $0 < \alpha < 1$. Denote by $\Lambda^{p,q}_{n,\alpha}$ the class of holomorphic functions $f \in H^1$ for which

$$|f|_{\Lambda^{p,q}_{n,\alpha}} = \max_{0 \le k \le n} \left(\int_{-\pi}^{\pi} \frac{1}{|t|^{1+\alpha q}} \left(\int_{-\pi}^{\pi} |f^{(k)}(e^{i(\theta+t)}) \right. \right.$$
$$\left. \left. - f^{(k)}(e^{i\theta})|^p \, d\theta \right)^{q/p} dt \right)^{1/q} < +\infty$$

c) *Functions with derivatives from the classes BMOA.* Finally, $BMOA_n$ will denote the class of holomorphic functions f for which we have the representation

$$f^{(n)}(z) = \frac{1}{2\pi} \int_{-\pi}^{\pi} \frac{e^{i\theta} + z}{e^{i\theta} - z} \psi(e^{i\theta}) \, d\theta, \qquad z \in \mathbf{D}, \ \psi \in L^\infty, \ n \ge 0;$$

in the class $BMOA_n$ we introduce the corresponding BMO norm, which makes $BMOA_n$ a Banach space.

THEOREM 1. *Let X be one of the spaces listed above and let $h \in H^\infty$. The operator*

$$T_h(f)(z) = \frac{1}{2\pi i} \int_\Gamma \frac{f(\varsigma)\overline{h(\varsigma)}}{\varsigma - z} \, d\varsigma, \qquad z \in \mathbf{D},$$

acts boundedly in X. Moreover,

$$\|T_h(f)\|_X \le \text{const} \, \|h\|_\infty \cdot \|f\|_X, \qquad f \in X.$$

In the case $X = H^p_n(\alpha)$, the proof of this theorem is based on the following auxiliary result. Let f be a function holomorphic in $\mathbf{D}$, $f(z) = \sum_0^{+\infty} a_k z^k$, and let $\alpha > -1$. Define the operator

$$D^\alpha f(z) = \sum_{k=0}^{+\infty} \frac{\Gamma(\alpha + 1 + k)}{\Gamma(\alpha + 1)\Gamma(k + 1)} a_k z^k.$$

LEMMA 1. *Let Φ be a continuous linear functional on $H^p(\alpha)$, $1 < p < +\infty$, and let $\delta_z(\varsigma) = 1/(1 - \varsigma z)$, $\varsigma, z \in \mathbf{D}$. The function $g(z) = \Phi(\delta_z)$ is holomorphic in D, $D^{\alpha+1}g \in H^q(\alpha)$, $1/p + 1/q = 1$, and for any $f \in H^p(\alpha)$*

$$\Phi(f) = \lim_{\rho \to 1-0} \frac{1}{2\pi} \int_{-\pi}^{\pi} f(\rho e^{i\theta}) g(e^{-i\theta}) \, d\theta. \tag{2}$$

Moreover, there exist positive numbers C_1 and C_2 such that

$$C_1 \|D^{\alpha+1}g\|_{H^q(\alpha)} \leq \|\Phi\| \leq C_2 \|D^{\alpha+1}g\|_{H^q(\alpha)}. \tag{3}$$

Conversely, by (2), every holomorphic function g such that $D^{\alpha+1}g \in H^q(\alpha)$ induces a continuous linear functional on $H^p(\alpha)$, for which the estimates (3) hold.

For $X = \Lambda_{n,\alpha}^{p,q}$ we use the arguments in [4].

Below we shall prove that the inclusion $h \in H^\infty$ in the well-known sense is necessary.

§**2.** Let ω be a function of the type of a modulus of continuity. The symbol $H^1(\omega, \alpha)$, $\alpha \geq 0$, will denote the class of functions f holomorphic in D for which

$$\|f\|_{H^1(\omega,\alpha)} = \int_{\mathbf{D}} |f(z)|\omega(1 - |z|)(1 - |z|)^{\alpha-1} \, d\sigma_2(z) < +\infty.$$

As above, let $H_n(\omega, \alpha)$ be the class of functions f such that $f^{(n)} \in H(\omega, \alpha)$ and let

$$\|f\|_{H_n^1(\omega,\alpha)} = \max_{0 \leq k \leq n} \left(\int_{\mathbf{D}} |f^{(k)}(z)|(1 - |z|)^{\alpha-1}\omega(1 - |z|) \, d\sigma_2(z) \right).$$

THEOREM 2. *Let $n > \alpha + 1$ and $h \in L^1(\Gamma)$. The following statements are equivalent:*

1) T_h *is a bounded operator in $H_n^1(\omega, \alpha)$.*

2) h *can be represented in the form*

$$h(e^{i\theta}) = \overline{h_1(e^{i\theta})} + h_2(e^{i\theta}), \qquad \theta \in [-\pi, \pi],$$

where $h_1 \in H_n(\omega, \alpha)$ and $h_2 \in H^\infty$.

For $n = \alpha + 1$ we have

THEOREM 3. *Let $h \in H^\infty$.*

1) *The following two statements are equivalent:*

   a) T_h *is a bounded operator in $H_n^1(\omega, n - 1)$;*

   b)

$$B_\omega(h) = \sup_{z \in \mathbf{D}} \left\{ |h'(z)| \frac{(1 - |z|)^2}{\omega(1 - |z|)} \int_{1-|z|}^{1} \frac{\omega(t)}{t^2} \, dt \right\} < +\infty.$$

2) *If $h \in H^\infty$ and $B_\omega(h) = +\infty$, then there exists a set M of second category in $H(\omega, n - 1)$, such that for any function $f \in M$ and*

$$J_n(f)(z) = \frac{1}{(n-1)!} \int_0^z (z - t)^{n-1} f(t) \, dt,$$

the function $T_h(J)n(f))$ does not belong to $H_n^1(\omega, n - 1)$.

We note that for $\omega(t) = t$, Theorem 3 was proved earlier by the author in [7].

Theorems 1–3 imply

THEOREM 4. *Let X coincide with one of the above spaces and, if $X = H_n^1(\omega, \alpha)$, assume that $n > \alpha + 1$.*

If $f \in X$ and J is an inner divisor of f, then fJ^{-1} belongs to X.

§3. Holomorphic functions with the absolute value of the boundary values from $W_n^p(\Gamma)$. The above results on division by an inner function show that if a function f is holomorphic in the unit disk and is "smooth" in the closed disk in one sense or another, then the function

$$Q_f(z) = \exp\left\{ \frac{1}{2\pi} \int_{-\pi}^{\pi} \frac{e^{i\theta} + z}{e^{i\theta} - z} \ln |f(e^{i\theta})| \, d\theta \right\}$$

also enjoys the same "smoothness" as f, at least in case the word "smoothness" is interpreted as membership of functions in the indicated classes.

Since the function Q_f is defined in terms of $|f(e^{i\theta})|$, the following question arises naturally. Let $h(x) \geq 0$, $x \in [-\pi, \pi]$, and $\log h \in L^1$, and let h be "smooth" in the above sense. What can be said about the smoothness of the outer function $Q_h \colon Q_h(e^{i\theta}) = h(\theta)$, $\theta \in [-\pi, \pi]$, in the closed disk? It turns out that under these circumstances, Q_h is, in general, twice as "bad" as h as far as smoothness is concerned, if we interpret the term "smoothness" as, for example, membership in some Lipschitz class α.

It has been established in [8] that if $h \in \operatorname{lip} \alpha$, $0 < \alpha < 1$, then $Q_h \in \operatorname{lip}(\overline{D}, \alpha/2)$ and the order cannot be improved. Khavin [9] subsequently extended this result to an arbitrary modulus of continuity $\omega(t)$ with the property that

$$\int_0 \frac{\omega(t)}{t} \, dt < +\infty$$

(cf. also [10] and [11]). Here we give a result from which it is obvious that an analogous phenomenon takes place in other scales of smooth functions.

THEOREM 5. *Assume that $h(x) \geq 0$, $\log h(x) \in L^1(-\pi, \pi)$, and $h \in W_{2n}^p(\Gamma)$, $1 \leq p \leq +\infty$. If*

$$Q(z) = \exp\left(\frac{1}{2\pi} \int_{-\pi}^{\pi} \frac{e^{i\theta} + z}{e^{i\theta} - z} \log h(\theta) \, d\theta \right), \qquad z \in \mathbf{D},$$

then $Q^{(n)} \in H^p$; moreover, there exists a function $h \in W_2^p(\Gamma)$ such that $Q' \notin H^{p_1}$ for any $p_1 > p$.

BIBLIOGRAPHY

1. Kenneth Hoffman, *Banach spaces of analytic functions*, Prentice-Hall, Englewood Cliffs, N.J., 1962.

2. V. V. Peller and S. V. Khrushchev, *Hankel operators, best approximations, and stationary Gaussian processes*, Uspekhi Mat. Nauk **37** (1982), no. 1 (223), 53–124; English transl. in Russian Math. Surveys **37** (1982).

3. V. P. Khavin, *On the factorization of analytic functions that are smooth up to the boundary*, Zap. Nauchn. Sem. Leningrad. Otdel. Mat. Inst. Steklov. (LOMI) **22** (1971), 202–205; English transl. in J. Soviet Math. **2** (1974), no. 2.

4. F. A. Shamoyan, *Division by an inner function in certain spaces of functions that are analytic in the disk*, Zap. Nauchn. Sem. Leningrad. Otdel. Mat. Inst. Steklov. (LOMI) **22** (1971), 206–208; English transl. in J. Soviet Math. **2** (1974), no. 2.

5. M. M. Dzhrbashyan, *On the canonical representation of functions that are meromorphic in the unit disk*, Akad. Nauk Armyan. SSR Dokl. **3** (1945), 3–9. (Russian)

6. ____, *On the problem of the representability of analytic functions*, Soobshch. Inst. Mat. Mekh. Akad. Nauk Armyan. SSR **2** (1948), 3–40. (Russian)

7. F. A. Shamoyan, *A class of operators connected with the factorization of analytic functions*, Zap. Nauchn. Sem. Leningrad. Otdel Mat. Inst. Steklov. (LOMI) **39** (1974), 200–205; English transl. in J. Soviet Math. **8** (1977), no. 1.

8. V. P. Khavin and F. A. Shamoyan, *Analytic functions with a Lipschitz modulus of the boundary values*, Zap. Nauchn. Sem. Leningrad. Otdel. Mat. Inst. Steklov. (LOMI) **19** (1970), 237–239; English transl. in Sem. Math. V. A. Steklov Math. Inst. Leningrad **19** (1970).

9. V. P. Khavin, *A generalization of the Privalov-Zygmund theorem on the modulus of continuity of the conjugate function*, Izv. Akad. Nauk Armyan. SSR Ser. Mat. **6** (1971), 252–258. (Russian)

10. N. A. Shirokov, *Ideals and factorization in algebras of analytic functions that are smooth up to the boundary*, Trudy Mat. Inst. Steklov. **130** (1978), 196–222; English transl. in Proc. Steklov Inst. Math. 1979, no. 4(130)

11. James E. Brennan, *Approximation in the mean by polynomials on non-Carathéodory domains*, Ark. Mat. **15** (1977), 117–168.

Translated by J. SZUCS

Amer. Math. Soc. Transl.
(2) Vol. **133**, 1986

Logarithmic Subharmonicity of a J-form
of an Analytic J-expanding Matrix-valued Function

UDC 517.5

V. È. KATSNEL'SON

This note is on the topic of the theory of J-expanding analytic matrix-valued functions (J-theory) founded by Potapov. The objects in J-theory are matrix-valued functions $W(z)$ that are analytic in the disk or in a half-plane and satisfy in their domain of analyticity the inequality

$$W(z)JW^*(z) - J \geq 0.$$

Here J is a matrix such that $J = J^*$ and $J^2 = I$, where I is the identity matrix.

The expression $WJW^* - J$, which is the defining structure of J-theory, is called the J-*form* of W. The matrix W, whose J-form is nonnegative Hermitian, is said to be J-*expanding*.

§**1.** Let O be an open set in the complex plane, and $X(z)$ a function defined for $z \in O$ whose values are $m \times m$ matrices.

DEFINITION. The function $X(z)$ is said to be *logarithmically subharmonic in O* if the function (scalar) $\varphi(z)X(z)\varphi^*(z)$ is subharmonic in O for any row vector-valued function $\varphi(z)$ analytic in O.

It follows from this definition that if $X(z)$ is an $m \times m$ matrix-valued function logarithmically subharmonic in O, and $G(z)$ is a $p \times m$ matrix-valued function analytic in O, then the matrix-valued function $Y(z) = G(z)X(z)G^*(z)$ is a $p \times p$ matrix-valued function logarithmically subharmonic in O.

It can be shown that a matrix-valued function $X(z)$ twice continuously differentiable in O is logarithmically subharmonic precisely when the following matrix

---

1980 *Mathematics Subject Classification* (1985 *Revision*). Primary 31A05; Secondary 30G30.

Translation of Dokl. Akad. Nauk Ukrain. SSR Ser. A **1982**, no. 7, 6–9; MR **84d**: 47047.

©1986 American Mathematical Society
0065-9290/86 $1.00 + $.25 per page

inequality holds:

$$\begin{bmatrix} \dfrac{\partial^2 X}{\partial z \partial \bar{z}}(z) & \dfrac{\partial X}{\partial z}(z) \\[2mm] \dfrac{\partial X}{\partial \bar{z}}(z) & X(z) \end{bmatrix} \geq 0 \qquad (z \in O),$$

where

$$\frac{\partial}{\partial z} = \frac{1}{2}\left(\frac{\partial}{\partial x} - i\frac{\partial}{\partial y} \right), \qquad \frac{\partial}{\partial \bar{z}} = \frac{1}{2}\left(\frac{\partial}{\partial x} + i\frac{\partial}{\partial y} \right).$$

The same is also true for nonsmooth functions, but only for locally integrable $X(z)$, if the derivatives are understood in the generalized sense.

Thus, the property of logarithmic subharmonicity is a local property for a function: if each point of O has a neighborhood in which $X(z)$ is logarithmically subharmonic, then $X(z)$ is a logarithmically subharmonic function on O.

If $X(z)$ is a logarithmically subharmonic matrix-valued function on O, then its norm $\|X(z)\|$ (the operator norm in Euclidean space) as well as its trace $\operatorname{tr} X(z)$ are scalar logarithmically subharmonic functions on O.

§**2.** LEMMA. *Let O be an open set in the complex plane that contains the half-planes* $\operatorname{Im} z > 0$ *and* $\operatorname{Im} z < 0$, *and perhaps some points of the real axis, and let $N(z)$ be an $m \times m$ matrix-valued function that is analytic in O and satisfies the inequalities*

$$\frac{N(z) - N^*(z)}{i} \geq 0 \quad (\operatorname{Im} z > 0), \qquad \frac{N(z) - N^*(z)}{i} \leq 0 \quad (\operatorname{Im} z < 0).$$

Suppose that for nonreal z the function $X(z)$ is given by

$$X(z) = \frac{N(z) - N^*(z)}{z - \bar{z}}, \tag{1}$$

and, at the points z of the real axis belonging to O, $X(z)$ is defined by continuity. Then $X(z)$ is logarithmically subharmonic on O.

PROOF. A matrix-valued function $C \geq 0$ independent of z is logarithmically subharmonic. The matrix-valued function $N(z)$ is representable in the form

$$N(z) = A + Bz + \int_{-\infty}^{\infty} \left(\frac{1}{\lambda - z} - \frac{\lambda}{1 + \lambda^2} \right) d\sigma(\lambda) \qquad (z \in O),$$

where $A = A^*$, $B \geq 0$, and $\sigma(\lambda)$ is an $m \times m$ matrix-valued function that is nondecreasing on $(-\infty, \infty)$ and such that

$$\int_{-\infty}^{\infty} (1 + \lambda^2)^{-1} \operatorname{tr} d\sigma(\lambda) < \infty,$$

and the set of points of increase of $\sigma(\lambda)$ does not intersect O. Consequently,

$$X(z) = B + \int_{-\infty}^{\infty} \frac{1}{\lambda - z}\, d\sigma(\lambda)\frac{1}{\lambda - \overline{z}} \qquad (z \in O).$$

If $\varphi(z)$ is any analytic row vector-valued function analytic in O, then (the integral is a limit of integral sums)

$$\varphi(z)X(z)\varphi^*(z) = \varphi(z)B\varphi^*(z) + \lim \sum \frac{\varphi(z)}{\lambda_k - z} C_k \frac{\varphi^*(z)}{\lambda_k - \overline{z}},$$

$$C_k = \sigma(\lambda_{k+1}) - \sigma(\lambda_k),$$

where the points λ_k do not lie in O and the limit is uniform with respect to z on each compact subset of O. Since each term in the integral sum is a function subharmonic in O, the limit of the sums is also a function subharmonic in O. The lemma is proved.

The next theorem (our main result) turns out to be useful in justifying various passages to limits in J-theory.

§3. THEOREM. *Let O be an open set in the complex plane that contains the half-planes $\mathrm{Im}\, z > 0$ and $\mathrm{Im}\, z < 0$, and perhaps some subset of the real axis, and let $W(z)$ be an $m \times m$ matrix-valued function that is analytic in O and satisfies the inequalities*

$$W(z)JW^*(z) - J \geq 0 \quad (\mathrm{Im}\, z > 0), \qquad W(z)JW^*(z) - J \leq 0 \quad (\mathrm{Im}\, z < 0).$$

Suppose that at nonreal points z the function $\Phi(z)$ is defined by[1]

$$\Phi(z) = i\frac{W(z)JW^*(z) - J}{z - \overline{z}}, \tag{2}$$

and at points z of the real axis belonging to O let $\Phi(z)$ be defined by continuity. Then $\Phi(z)$ is a matrix-valued function logarithmically subharmonic on O.

PROOF. Let z_0 be an arbitrary point of O. Since logarithmic subharmonicity is a local property, it suffices to show that $\Phi(z)$ is logarithmically subharmonic near z_0. Choose $\alpha, \alpha = \overline{\alpha}$, such that the matrix $W(z_0) - e^{i\alpha}I$ is invertible. Let

$$N(z) = i(W(z) - e^{i\alpha}I)^{-1} \cdot (W(z) + e^{i\alpha}I)J.$$

The function $N(z)$ is analytic near z_0 and has in O only isolated singularities: poles. Since

$$\frac{N(z) - N^*(z)}{2i} = (W(z) - e^{i\alpha}I)^{-1}(W(z)JW(z)^* - J)(W(z) - e^{i\alpha}I)^{-1*},$$

it follows that

$$\frac{N(z) - N^*(z)}{i} \geq 0 \quad (\mathrm{Im}\, z > 0), \qquad \frac{N(z) - N^*(z)}{i} \leq 0 \quad (\mathrm{Im}\, z < 0),$$

---

[1] The expression on the right-hand side of (2) is called a *J-form*, just like the expression $W(z)JW^*(z) - J$.

and hence the singularities of $N(z)$ in the upper and lower half-planes are removable (it will be assumed that they have been removed). According to the lemma, the function $X(z)$ defined in (1) is logarithmically subharmonic in the union of the upper and lower half-planes and some neighborhood of z_0 (it is only necessary to mention the neighborhood of z_0 in the case when this point is real). Since $\Phi(z) = G(z)X(z)G^*(z)$, where $G(z) = W(z) - e^{i\alpha}I$, it follows that $\Phi(z)$ is logarithmically subharmonic near z_0.

The theorem is proved.

Translated by H. H. McFADEN

Amer. Math. Soc. Transl.
(2) Vol. **133**, 1986

Meromorphic Solutions of Equations
of Briot-Bouquet Type

UDC 519.925.6

A. È. EREMENKO

An important problem in the analytic theory of differential equations (d.e.) is the study of single-valued solutions. There are a large number of papers devoted to algebraic d.e. whose general solution is single-valued in $\mathbf{C}$. A more complicated problem is the study of differential equations possessing particular meromorphic solutions.

In 1920 Malmquist completely investigated the algebraic d.e. of the first order having a meromorphic particular solution. Very little is known about the particular solutions of higher-order d.e. It is therefore natural to restrict oneself to some simple class of equations.

This note is devoted to d.e. of Briot-Bouquet type

$$P(y^{(k)}, y) = P_0(y)(y^{(k)})^m + \cdots + P_m(y) = 0, \tag{1}$$

where P is a polynomial in two variables. For $k = 1$ these are the classical Briot-Bouquet equations, which can have meromorphic (in $\mathbf{C}$) solutions of the following three types: elliptic functions, rational functions, and rational functions of $\exp(az)$, $a \in \mathbf{C}$. We shall call them *functions of class W*. This class arises when studying algebraic addition theorems as well.

Picard [1] proved that for $k = 2$ any meromorphic solution of (1) also belongs to the class W. Hille [2]–[3] studied the order of growth of entire and meromorphic solutions of some types of equations (1), as well as the properties of multiple-valued solutions of these equations.

It seems plausible that any meromorphic function satisfying (1) belongs to the class W. We shall prove this in two cases: when the genus of the polynomial P

1980 *Mathematics Subject Classification* (1985 *Revision*). Primary 34A20; Secondary 30D35.

Translation of Teor. Funktsiĭ, Funktsional. Anal. i Prilozhen. Vyp. **38** (1982), 48–56; MR **80h**:34013.

©1986 American Mathematical Society
0065-9290/86 $1.00 + $.25 per page

equals 1 (Theorem 2) and when k is odd and the solution is not entire (Theorem 3). Theorem 1 provides easily verifiable necessary conditions for the existence of entire and meromorphic solutions of d.e. (1).

We make use of the simplest facts and standard notation of Nevanlinna theory [4], [5] and the theory of algebraic functions [6], without special notice.

We shall assume further that the polynomial P in (1) is irreducible, as a polynomial in two variables; this is no loss of generality when studying the properties of the solutions. To equation (1) there corresponds the algebraic function $p(y)$, determined by the relation $P(p, y) = 0$. Let F be the Riemann surface of this function and $\overline{\mathbf{C}}_y$ the covering plane. We shall represent F as the set of pairs of elements (p, y) satisfying $P(p, y) = 0$.

Each meromorphic solution $y(z)$ of (1) induces a covering map $f \colon \mathbf{C}_z \to F$. This map uniformizes the Riemann surface F by means of the pair of functions $(y^{(k)}, y)$. We shall need the following fact.

THEOREM A (PICARD). *Assume that the meromorphic map $f \colon \mathbf{C} \to F$ is not rational. Then the genus of the surface F equals 0 or 1. If the genus equals 1, then every point of F has infinitely many inverse images in $\mathbf{C}$. If the genus of F equals 0, then every point of F, with at most two exceptions, has infinitely many inverse images.*

Usually the third statement of this theorem is called Picard's theorem. The first statement was proved by Picard in 1887 and was immediately afterwards applied to equation (1) in [1]. In the present form Theorem A can be found in [7] (p. 206 and the remark on p. 193).

From the first statement of Theorem A it follows that if the d.e. (1) has a meromorphic solution, then the genus of F equals 0 or 1 [1].

To any point $M \in F$ lying above $y = \infty$, there corresponds the asymptotic relation

$$p(y) = (a + o(1))y^q, \qquad y \to \infty, \ a = a(M) \in \mathbf{C} \tag{2}$$

where $q = q(M)$ is a rational number. The letter M without index will always denote a point of F, which is projected into $y = \infty$.

THEOREM 1. *Assume that the d.e. (1) has a transcendental meromorphic solution $y(z)$. Then $P_0 \equiv \mathrm{const}$, and the numbers q in (2) have the following properties:*

1°. *$q(M) = 1$ for at most two points M; for all the other points $q(M) = 1 + k/n$, where n is a natural number.*

2°. *For the function $y(z)$ to be entire, it is necessary and sufficient that the surface F have genus 0 and have at most two points above $y = \infty$; furthermore, $q = 1$ in each of these points.*

3°. *If $y(z)$ has a pole of order n, then for one of the points M we have $q(M) = 1 + k/n$; if for one of the points M we have $q(M) = 1 + k/n$, then $y(z)$ has infinitely many poles of order n.*

The numbers $q(M)$ in (2) are computed directly from the polynomial P with the help of the Newton diagram ([6], §38). In the case under consideration, the Newton diagram is constructed as follows. To each term $a_{ij}p^i y^j$ of the polynomial $P(p, y)$ we put in correspondence the point in the first quadrant with coordinates i, j. We add moreover the points $(0,0)$ and $(m,0)$, and consider the convex hull of the set thus obtained. Consider all the sides of this polygon, with the exception of the vertical segments and the segment of the axis i. The slopes of these sides, taken with the opposite sign, are precisely the numbers $q(M)$ in (2). If (1) has a meromorphic solution, then, according to Theorem 1, $\deg P_0 = 0$ and $q(M) \leq 1+k$ for all points M. Therefore the Newton diagram lies no higher than the line $j = -(1 + k)(i - m)$, whence we obtain

$$\deg P_i \leq i(1 + k), \qquad i = 0, \ldots, m, \tag{3}$$

If d.e. (1) has an entire solution, then $q(M) = 1$ for all M, according to Theorem 1. From the Newton diagram we obtain

$$\deg P_i \leq i, \quad i = 0, \ldots, m; \qquad \deg P_m = m. \tag{4}$$

In most theorems of Hille [2], [3] conditions (3) and (4) are assumed to hold a priori. Theorem 1 shows that these assumptions are superfluous. From Theorem 5 of [3] it follows that every entire solution of equation (1) is a function of exponential type. This is also easily derived from the Wiman-Valiron theory ([4], Chapter 5) without using Theorem 1.

PROOF OF THEOREM 1. Assume $P_0 \not\equiv$ const. The algebraic function $p(y)$ has in this case a pole $M_0 \in F$, and M_0 is projected onto the point $a \in \mathbf{C}_y$, one of the roots of the polynomial P_0. It is apparent that M_0 does not belong to the image of the plane $\mathbf{C}$ under the map f. Consider a sufficiently small ε-neighborhood $V \subset F$ of M_0 and denote by D a component of the inverse image $f^{-1}(V)$. The set $\overline{V} \backslash M_0$ is not compact and consequently the domain D is unbounded. Consider in D the analytic function $w(z) = (y(z) - a)^{-1}$, where $y(z)$ is a meromorphic solution of equation (1). We have $|w| = \varepsilon^{-1}$ on ∂D and $|w| > \varepsilon^{-1}$ in D; therefore $w(z)$ is unbounded in D. Since M_0 is a pole of the algebraic function $p(y)$, we have

$$y^{(k)}(z) \sim \text{const.} \cdot w^{\varkappa}(z), \qquad z \in D, \ \kappa > 0, \ |w| \to \infty. \tag{5}$$

Let us show that (5) leads to a contradiction. To this end, we differentiate the relation $y(z) = w^{-1}(z) + a$ with respect to z a total of k times; we obtain

$$y^{(k)} = \frac{1}{w} Q\left(\frac{w'}{w}, \frac{w''}{w}, \ldots, \frac{w^{(k)}}{w}\right), \tag{6}$$

where Q is a polynomial in k variables. Further we make use of the theory developed in §17 of [8]. Let $G \subset \mathbf{C}_t$ be the total image of the domain D under the map $t = \log z$. The domain G meets every line $\operatorname{Re} t = x$ for x large enough. Consider in G the analytic function $v(t) = w(e^t)$. It is easily seen that this function is bounded on ∂G and on the intersection of each vertical line with G, but $v(t)$ is unbounded in G. Set

$$S(x) = \max_{\mathrm{Re}\, t=x,\ t\in G} |v(t)|, \qquad L(x) = \frac{S'(x)}{S(x)}.$$

(The maximum in the definiton of $S(x)$ is always attained at some point $\varsigma = \varsigma(x) \in G$; in the definition of $L(x)$ and also in analogous cases the right-hand derivative is considered.) According to Theorem 1.4.17 of [8], we have (formula (5.4.17))

$$\lim_{\substack{x\to\infty \\ x\notin E}} \frac{1}{L^j(x)} \frac{v^{(j)}(\varsigma)}{S(x)} = 1, \qquad j = 1, 2, \dots, \tag{7}$$

where E is a set of finite measure on the half-axis $x > 0$. (In Chapter III of [8] all the arguments are made for the case when v is defined in the half-plane $\mathrm{Re}\, t > 0$. The analysis of the arguments which precede Theorem 1.4.17 shows that they are all valid in our case as well. One needs only verify that the small disk (5.2.17) in the statement of Theorem 1.2.17 is contained in the domain G. This follows from (9.2.17). Further arguments from §17 of [8] can be taken over word for word.)

Let us deduce from (7) a similar relation for the function w. We put

$$M(r,w) = \max\{|w(z)|\colon z \in D,\ |z| = r\}, \qquad K(r) = rM'(r,w)/M(r,w).$$

Taking into account the fact that $v(t) = w(\exp t)$ and arguing as in the proof of Theorem 1.10.17 from [8], we obtain from (7)

$$|w^{(j)}(s)| = (1 + o(1))\frac{K^{(j)}(r)}{r^j} M(r,w), \qquad j = 1, 2, \dots, \tag{8}$$

where $r = |s| \to \infty$ outside a set of finite logarithmic measure; s is a point where $|w(s)| = M(r)$. According to Lemma 1.5.2 from [8], for any $a > 0$ we have

$$(K(r))^a = o(M(r,w)), \tag{9}$$

when $r \to \infty$ outside a set of finite logarithmic measure. Taking (8) and (9) into account, we obtain from (6) that $y^{(k)}(s) \to 0$ for some sequences (s_j), $|s_j| \to \infty$. But $w(s_j) = M(|s_j|,w) \to \infty$, which contradicts (5). This contradiction shows that $P_0 \equiv \mathrm{const}$.

In order to prove the remaining statements of Theorem 1, let us consider a point $M \in F$ which projects onto $y = \infty$. If M has at least one inverse image $z_0 \in \mathbf{C}$ under the map $f\colon \mathbf{C} \to F$, then z_0 is a pole of the function $y(z)$. We have $y(z) \sim \mathrm{const}(z - z_0)^{-n}$, where n is the order of the pole. Then $y^{(k)}(z) \sim \mathrm{const}(z - z_0)^{-n-k}$ as $z \to z_0$. Consequently, $q(M) = 1 + k/n$ in (2).

Assume now that the point M has a finite number of inverse images in $\mathbf{C}$. Let us show that the number $q(M)$ in this case equals 1. Consider a neighborhood of the point M, i.e. a component V of the set $F \cap \{y\colon |y| > \varepsilon^{-1}\}$ such that $M \in V$. If ε is sufficiently small, then the unbounded component D of the inverse image $f^{-1}(V)$ contains no poles of the function $y(z)$; moreover, $|y| = \varepsilon^{-1}$ on ∂D and

$y(z)$ is unbounded in D. Applying (8) and (9) to $y(z)$ instead of $w(z)$, we get $y^{(k)}(s) = O((y(s))^{1+\eta})$ as $s \to \infty$ for any $\eta > 0$; consequently $q \le 1$ in (2).

If $q(M) < 1$, we apply (8) once again, taking into account that $K(r)$ is not decreasing:

$$M^q(r, y) = (|y(s)|)^q \sim \operatorname{const}|y^{(k)}(s)| \ge \operatorname{const} \cdot r^{-k} M(r, y),$$

$$M(r, y) = O(r^\alpha), \qquad r \to \infty, \ \alpha > 0. \tag{10}$$

According to Theorem A, the surface F has genus 0, whence F is birationally equivalent to the sphere $\overline{\mathbf{C}}_t$. This means that there exists a rational function $t = R(p, y)$ which maps F bijectively onto $\overline{\mathbf{C}}_t$. We can assume that the point $M \in F$ is taken into ∞ under this map. Then the function

$$t(z) = R(y^{(k)}(z), y(z)) \tag{11}$$

is entire, and from (11) and (10) we obtain that $|t(z)| \le \operatorname{const}|z|^\beta$, $\beta > 0$. Consequently, $t(z)$ is a polynomial. Evidently, the numerator and the denominator of $R(p, y)$ are relatively prime to $P(p, y)$. Therefore, after eliminating $y^{(k)}$ from (1) and (11) we obtain $Q(y(z), z) = 0$, where $Q \not\equiv 0$ is a polynomial. Then $y(z)$ cannot be a transcendental function.

We have proved that for a point $M \in F$ which projects onto $y = \infty$ there are two mutually exclusive possibilities: either $q(M) = 1$ and the point M has no inverse images in $\mathbf{C}$, or $q(M) = 1 + k/n$ and M has infinitely many inverse images, which are poles of order n of $y(z)$. Together with Theorem A this provides $1°$, $2°$ and $3°$ of Theorem 1.

THEOREM 2. *If F is a surface of genus* 1, *then every meromorphic solution of equation* (1) *is an elliptic function.*

PROOF. The universal covering $\mathbf{C} \to F$ of a surface of genus 1 is realized by means of a pair of elliptic functions Φ_1 and Φ_2 with the same period lattice. If (1) has a meromorphic solution $y(z)$, then for a pair of functions $y(z)$, $y^{(k)}(z)$ which provide the covering $f \colon \mathbf{C} \to F$, the following holds:

$$y(z) = \Phi_1(\varphi(z)), \qquad y^{(k)}(z) = \Phi_2(\varphi(z)), \tag{12}$$

where φ is an entire function (this fact was observed by Picard [1]). We must show that $\varphi(z)$ is a linear function. Assuming this is not the case, let us differentiate the first relation (12) k times and subtract the second relation (12), to obtain

$$\Phi_1^{(k)}(\varphi(z))g_k(z) + \cdots + \Phi_1'(\varphi(z))g_1(z) - \Phi_2(\varphi(z)) \equiv 0.$$

Here the functions $g_j(z)$ are the products of derivatives of $\varphi(z)$ and constant coefficients. For instance, $g_k(z) = (\varphi'(z))^k \not\equiv \operatorname{const}$, if φ is not linear. The functions $\Phi_1, \ldots, \Phi_1^{(k)}$ are linearly independent over $\mathbf{C}$. The function Φ_2 either is linearly independent of $\Phi_1, \ldots, \Phi_1^{(k)}$ or is a linear combination of them, with constant coefficients. In the second case we put this linear combination instead of Φ_2 and combine the terms containing equal elliptic functions. We obtain

$$\psi_1(\varphi(z))h_1(z) + \cdots + \psi_n(\varphi(z))h_n(z) \equiv 0, \tag{13}$$

where the number n satisfies the condition $1 \leq n \leq k + 1$, the ψ_j are linearly independent (over $\mathbf{C}$) elliptic functions with the same period lattice Γ, and the h_j are entire functions which are rational functions of $\varphi(z)$ and its derivatives. At least one of the functions h_j is not a constant.

Let us show that a relation (13) with such properties is impossible. Let X_0 be the set of Valiron exceptional values of $\varphi(z)$. This means that for $a \notin X_0$

$$N(r, a, \varphi) \sim T(r, \varphi), \qquad r \to \infty. \tag{14}$$

Let $X_1, X_2, \ldots$ be the translation of the set X_0 by all the periods of the lattice Γ. Set $X = \bigcup_0^\infty X_j$. It is known that X_0 is a set of plane measure 0 ([5], p. 151). Since the ψ_j are linearly independent over $\mathbf{C}$, there exist points $\varsigma_1, \ldots, \varsigma_n$ in $\mathbf{C} \backslash X$ such that

$$\det \|\psi_i(\varsigma_j)\| \neq 0. \tag{15}$$

Evidently, the ς_j are pairwise noncongruent $\bmod \Gamma$. Let us fix an arbitrary j, $1 \leq j \leq n$. The function

$$H_j(z) = \psi(\varsigma_j)h_1(z) + \cdots + \psi_n(\varsigma_j)h_n(z) \tag{16}$$

is a polynomial in the derivatives of φ. By the lemma about the logarithmic derivative ([5], p. 122) we have

$$T(r, H_j) \leq LT(r, \varphi), \qquad r \notin E, \tag{17}$$

where E is a set of finite length and L is a sufficiently large natural number. For each $j = 1, \ldots, n$ let us choose a set of $L + 1$ different points ς_{ij}, $1 \leq i \leq L + 1$, which are congruent to $\varsigma_j \bmod \Gamma$. Since $\varsigma_j \notin X$, then $\varsigma_{ij} \notin X$ and by virtue of (14) we have

$$N(r, \varsigma_{ij}, \varphi) \sim T(r, \varphi), \qquad j = 1, \ldots, n, \; i = 1, \ldots, L + 1. \tag{18}$$

Consider a sequence $(z_{lj})_{l=1}^\infty = Z_j$ of points where the function has values from the set $\{\varsigma_{ij}\}_{i=1}^{L+1}$. Let $N(r, Z_j)$ be the counting function of the sequence Z_j. From (18) we get $N(r, Z_j) \sim (L + 1)T(r, \varphi)$, $r \to \infty$. Substituting any point $z_{lj} \in Z_j$ into (13) and taking into account the fact that $\varphi(z_{lj}) \equiv \varsigma_j \pmod{\Gamma}$ and also (16), we obtain $H_j(z_{lj}) = 0$. Consequently,

$$N(r, 0, H_j) \geq (L + 1)T(r, \varphi).$$

Together with (17) and the Jensen inequality, this gives $H_j(z) \equiv 0$. Equations (16) turn into an homogeneous system of algebraic equations with nonzero determinant by (15). Consequently, all functions $h_j(z) \equiv 0$. This contradiction proves the theorem.

In what follows we shall need the following fact.

LEMMA. *Let k be odd. Then the set of meromorphic functions with poles at 0 satisfying equation (1) is finite.*

PROOF. Assume that one of the meromorphic solutions of (1) has a pole at 0 of order n. Then, by Theorem 1, for one of the points $M \in F$ over $y = \infty$ we have $q(M) = 1 + k/n$.

In a neighborhood of M the algebraic function corresponding to equation (1) has a Puiseux expansion $p(y) = y^q(a_0 + a_1 y^{-1/n} + \cdots)$. The meromorphic solution in a neighborhood of 0 has the form

$$y(z) = c_0 z^{-n} + c_1 z^{-n+1} + \cdots, \qquad c_0 \neq 0. \tag{19}$$

This power series must formally satisfy the equation obtained from the Puiseux equation

$$y^{(k)}(z) = a_0 y^{1+k/n} + \cdots, \qquad a_0 \neq 0. \tag{20}$$

The coefficients in the right-hand side of (20) are determined by (1). We shall show that c_0 is determined to within a finite set, and that the c_j, for $j \geq 1$, are uniquely determined if c_0 is given. We have

$$y^{(k)}(z) = (-1)^k \left[\frac{(k+n-1)!}{(n-1)!} c_0 z^{-n-k} + \frac{(k+n-2)!}{(n-2)!} c_1 z^{-n-k+1} \right.$$
$$\left. + \cdots + k! c_{n-1} z^{-k-1} \right]$$
$$+ k! c_{n+k} + \frac{(k+1)!}{1!} c_{n+k+1} z + \frac{(k+2)!}{2!} c_{n+k+2} z^2 + \cdots;$$
$$y^{1+k/n} = z^{-k-n} \left(c_0^{1+k/n} + \left((1+k/n) c_0^{k/n} c_1 + (\cdots)_1 \right) z \right.$$
$$+ ((1+k/n)c_0^{k/n} c_2 + (\cdots)_2)z^2 + \cdots$$
$$\left. + ((1+k/n)c_0^{k/n} c_j + (\cdots)_j)z^j + \cdots \right).$$

In the second formula the symbol $(\cdots)_j$ denotes a finite sum of products of the coefficients of the series (19), which contains no coefficient c_i with index $i \geq j$. Comparing the coefficients of z^{-k-n} in (20), we get

$$(-1)^k \frac{(k+n-1)!}{(n-1)!} c_0 = a_0 c_0^{1+k/n}.$$

This equation with respect to c_0 has a finite number of nonzero roots. We have

$$a_0 c_0^{k/n} = (-1)^k \frac{(k+n-1)!}{(n-1)!}. \tag{21}$$

Further we obtain

$$(-1)^k \frac{(k+n-2)!}{(n-2)!} c_1 = a_0 c_0^{k/n} \left(1 + \frac{k}{n} \right) c_1 + (\cdots)_1. \tag{22}$$

Substituting here the value of $a_0 c_0^{k/n}$ determined from (21), we can see that the coefficient of c_1 is different from 0, because

$$(-1)^k \frac{(k+n-2)!}{(n-2)!} \neq (-1)^k \frac{(k+n-1)!}{(n-1)!} \frac{k+n}{n}.$$

Thus, c_1 is uniquely determined from (22). The situation is analogous for all the coefficients c_j, $j < n+k$. For the coefficient c_{n+k+j}, $j \geq 0$, we have

$$\frac{(k+j)!}{j!} c_{n+k+j} = a_0 c_0^{k/n} \frac{n+k}{n} c_{n+k+j} + (\cdots)_{n+k+j}.$$

Once again we put the value of $a_0 c_0^{k/n}$ from (21) and conclude that the coefficient of c_{n+k+j} equals

$$\frac{(k+j)!}{j!} - (-1)^k \frac{(k+n)!}{n!}. \tag{23}$$

Since k is odd, this coefficient is different from 0 for all j. Thus, all c_j are uniquely determined if c_0 is given, and the lemma is proved.

REMARK. From (23) we see that if k is even, the coefficient c_{2n+k} cannot be determined from equation (20). Therefore, the reasoning on p. 274 of [3] seems unconvincing for k even.

THEOREM 3. *Assume that F is of genus 0, k is odd and a meromorphic solution of equation* (1) *has at least one pole. Then this solution belongs to the class W.*

PROOF. Assume that $y(z)$ is a transcendental function. By Theorem 1, $3°$, $y(z)$ has infinitely many poles z_j, $j = 1, 2, 3, \ldots$. The functions $y(z - z_j)$, $j = 1, 2, \ldots$, satisfy the assumptions of the lemma, and therefore some of them coincide. Consequently, $y(z)$ is a periodic function. We shall assume (with no loss of generality) that the least period, in absolute value, equals $2\pi i$.

Consider the strip $D = \{z \colon 0 \le \operatorname{Im} z < 2\pi\}$.

First case. $y(z)$ is bounded in $D \cap \{z \colon |\operatorname{Re} z| > A\}$ for some $A > 0$. In particular, $y(z)$ has a finite number of poles in D. It is easy to see that in this case $y(z)$ takes each complex value finitely many times in D. Since $y(z)$ is periodic, the function $R(z) = y(\ln z)$ is single-valued in $\{z \colon 0 < |z| < \infty\}$. This function takes each value finitely many times; hence it is rational. Consequently, $y(z) = R(\exp z) \in W$.

Second case. $y(z)$ has infinitely many poles in D. Repeating the reasoning given at the beginning of the proof and using the lemma, we conclude that the function $y(z)$ is elliptic.

Third case. $y(z)$ has a finite number of poles in D and is unbounded in $D \cap \{z \colon |\operatorname{Re} z| > A\}$ for any $A > 0$. Assume, for instance, that $y(z)$ is unbounded in $\{z \colon \operatorname{Re} z > A\}$ for any $A > 0$. Since $y(z)$ is periodic with period $2\pi i$, and the half-plane $\{z \colon \operatorname{Re} z > A\}$ contains no poles of $y(z)$, this function belongs to the class Π of Strelitz ([8], §17). For $y(z)$ the following relation holds, analogous to (7):

$$\lim_{\substack{x \to \infty \\ x \notin E}} L^{-k}(x) \frac{|y^{(k)}(\varsigma)|}{S(x, y)} = 1. \tag{24}$$

Here

$$|y(\varsigma)| = \max_{\operatorname{Re} z = x} |y(z)| = S(x, y), \qquad L(x) = \frac{S'(x, y)}{S(x, y)}.$$

E is a set of finite measure. Our goal is to prove that $L(x) = O(1)$, $x \to \infty$. Consider the component $G \subset \{z \colon \operatorname{Re} z > 0\}$ of the set $\{z \colon |y(z)| > \varepsilon^{-1}\}$, where $\varepsilon > 0$ is so small that G contains no poles of the function $y(z)$. The component G is mapped under $f \colon \mathbf{C} \to F$ into a neighborhood of some point $M \in F$, which is projected onto $y = \infty$, and M has no inverse images in $\mathbf{C}$. Reasoning as in the proof of Theorem 1, we conclude that $q(M) = 1$. Therefore, by (2) in G we have

$$|y^{(k)}(z)| \sim |a(M)| \cdot |y(z)|, \qquad y(z) \to \infty, \quad a(M) \in \mathbf{C}\backslash\{0\}.$$

There are a finite number of points M which are projected onto ∞; hence

$$|y^{(k)}(z)| \le a_0|y(z)| \quad \text{for } |y(z)| > A \tag{25}$$

for some constants a_0 and A, independent of the choice of the component G. From (24) and (25) it follows that $L(x) = O(1)$, $x \to \infty$.

Since $L(x) = (d/dx)\ln S(x,y)$, we have $\ln S(x,y) = O(x)$. From this fact and from the periodicity of the function y we easily obtain $m(r,y) = O(r)$, $r \to \infty$, where $m(r,y)$ is the Nevanlinna approximation function. It is also evident that $N(r,y) = O(r)$, $r \to \infty$, since $y(z)$ has period $2\pi i$, and the number of poles in D is finite. From this fact and from (26) we conclude that $T(r,y) = O(r)$, $r \to \infty$. Now by the first fundamental theorem of Nevanlinna, $N(r,a,y) = O(r)$, $r \to \infty$, for all $a \in \mathbf{C}$. By the periodicity, each value is taken in D finitely many times. We have again the first case.

The theorem is proved.

REFERENCES

1. E. Picard, *Sur une propriété des fonctions uniformes d'une variable et sur une classe d'équations différentielles*, C. R. Acad. Sci. Paris **91** (1880), 1058–1061.

2. Einar Hille, a) *Remarks on Briot-Bouquet differential equations*. I, Comment. Math. Special Issue **1** (1978), 119–132.

b) *Some remarks on Briot-Bouquet differential equations*. II, J. Math. Anal. Appl. **65** (1978), 572–585.

3. ——, *Higher order Briot-Bouquet differential equations*, Ark. Mat. **16** (1978), 271–286.

4. Hans Wittich, *Neuere Untersuchungen über eindeutige analytische Funktionen*, Springer-Verlag, 1955.

5. A. A. Gol'dberg and I. V. Ostrovskiĭ, *Distribution of values of meromorphic functions*, "Nauka", Moscow, 1970. (Russian)

6. N. G. Chebotarev, *Theory of algebraic functions*, OGIZ, Moscow, 1948. (Russian)

7. Phillip Griffiths and James King, *Nevanlinna theory and holomorphic mappings between algebraic varieties*, Acta Math. **130** (1973), 145–220.

8. Sh. I. Strelits [Shlomo Strelitz], *Asymptotic properties of analytic solutions of differential equations*, "Mintis", Vilnius, 1972. (Russian)

Translated by J. J. TOLOSA

Amer. Math. Soc. Transl.
(2) Vol. **133**, 1986

The Generalized Robin Problem and a Variational Method for Solving It

L. A. MARKUSHEVICH

The well-known Robin problem of using given charges on conductors (whose distribution is also given) to find the potential of each of them can be formulated as follows. Let Ω be a multiply connected region whose boundary components (conductors) are $\Sigma_0, \Sigma_1, \ldots, \Sigma_N$, where all the surfaces Σ_k with $k > 0$ are contained inside Σ_0. Suppose that $u(x)$ (the potential field) satisfies the equation $\Delta u = 0$, $x = (x_1, x_2, x_3) \in \Omega$, and the boundary conditions

$$u(x)|_{\Sigma_i} = \alpha_i = \text{const}, \qquad i = 0, 1, \ldots, N, \quad \alpha_0 = 0; \tag{1}$$

$$\int_{\Sigma_i} \frac{\partial u}{\partial n} = q_i, \qquad i = 1, \ldots, N. \tag{2}$$

From the given numbers $q_1, \ldots, q_N$ it is required to find either the function $u(x)$ itself or the unknowns $\alpha_1, \ldots, \alpha_N$. The Robin problem can be generalized in a natural way. Let

$$Au = -\sum_{k,j=1}^{n} \frac{\partial}{\partial x_j} \left(a_{jk} \frac{\partial u}{\partial x_k} \right) + c(x)u, \tag{3}$$

$$x = (x_1, \ldots, x_n), \quad a_{jk} = a_{kj}, \quad c(x) \geq 0,$$

be a uniformly elliptic operator in Ω. We assume that $a_{jk}(x) \in C^1(\Omega) \cap C^0(\Omega)$, $c(x) \in C^0(\Omega)$, and the boundary of Ω is sufficiently smooth. The following problem $R_{i_0, i_1, \ldots, i_s}$ is the generalized Robin problem, where $i_0 = 0 < i_1 < \cdots < i_s \leq N$: find a solution of the equation

$$Au = f(x), \qquad x \in \Omega, \ f(x) \in C^0(\overline{\Omega}), \tag{4}$$

1980 *Mathematics Subject Classification* (1985 *Revision*). Primary 35J20, 35J25; Secondary 49D15.

Translation of Trudy Moskov. Ènerget. Inst. Vyp. 412 (1979), 93–97; MR **81k**:35047.

©1986 American Mathematical Society
0065-9290/86 $1.00 + $.25 per page

satisfying the condition

$$\int_{\Sigma_{j_m}} \frac{\partial u}{\partial \nu}\, ds = q_{j_m}, \qquad m = s+1, \ldots, N, \ j_m \neq i_k, \ j_m < j_{m+1} \qquad (5)$$

($\partial u/\partial \nu$ is the derivative with respect to the conormal) and condition (1), where $\alpha_{i_0} = 0$; $\alpha_{i_1}, \ldots, \alpha_{i_s}$ and $q_{j_{s+1}}, \ldots, q_{j_N}$ are given numbers.

Obviously $R_{0,1,\ldots,N}$ is a special Dirichlet problem for equation (4). Problem R_0 is the Robin problem formulated above when $A = -\Delta$, $f(x) = 0$, and $n = 3$. Problem $R_{i_0,i_1,\ldots,i_s}$ ($s > 0$) is equivalent to finding the potential function from the given potentials of the conductors $\Sigma_0, \Sigma_{i_1}, \ldots, \Sigma_{i_s}$ and the given charges on the conductors $\Sigma_{j_{s+1}}, \ldots, \Sigma_{j_N}$.

We prove the possibility of finding a solution of the problem $R_{i_0,i_1,\ldots,i_s}$ by a variational method.

§1. Positive definiteness of the operator A

Let $H[i_0, i_1, \ldots, i_s; \alpha_{i_0}, \ldots, \alpha_{i_s}]$ be the set of functions in $C^2(\Omega) \cap C^1(\overline{\Omega})$ satisfying condition (1) with the given numbers $\alpha_{i_0} = 0, \ldots, \alpha_{i_s}$, and let

$$H[i_0, \ldots, i_s; \alpha_{i_0}, \ldots, \alpha_{i_s}; q_{j_{s+1}}, \ldots, q_{j_N}]$$

be the set of functions in $H[i_0, \ldots, i_s; \alpha_{i_0}, \ldots, \alpha_{i_s}]$ satisfying also condition (5). Obviously, $H_0 = H[0; 0]$, $H_s = H[0, \ldots, i_s; 0, \ldots, 0]$, and $H_s^0 = H[0, \ldots, i_s; 0, \ldots, 0; 0, \ldots, 0]$ are linear spaces. It is easy to see that the operator A is symmetric and positive on H_s^0. Indeed, A satisfies Green's formula

$$(Au, v) = \int_\Omega \left(\sum a_{jk} \frac{\partial u}{\partial x_k} \frac{\partial u}{\partial x_j} + cuv \right) dx - \int_{\partial\Omega} v \frac{du}{d\nu}\, ds. \qquad (6)$$

If both functions u and v belong to H_s^0, then the integral over the boundary on the right-hand side of (6) disappears. For such functions we have $(Au, v) = (u, Av)$, i.e., A is symmetric in H_s^0. Further, setting $u = v$ in (6), we find that

$$(Au, u) \geq \gamma(\nabla u, \nabla u) + (cu, u) \geq \gamma(\nabla u, \nabla u), \quad \gamma > 0,$$

for any $u \in H_s^0$. In a way analogous to that in [1] (Chapter IV, §24, but for other boundary conditions and for a simply connected region) it is possible to get an estimate $(\nabla u, \nabla u) \geq j_0(u, u)$, where $j_0 > 0$ and $u \in H_0 \supset H_s^0$. This will imply that A is positive-definite in H_s^0.

§2. Reduction of problem $R_{i_0,\ldots,i_s}$ to a variational problem

Suppose first that the boundary conditions of the problem are homogeneous: $\alpha_{i_0} = 0 = \alpha_{i_1} = \cdots = \alpha_{i_s}$; $q_{j_{s+1}} = \cdots = q_{j_N} = 0$. Since A is positive-definite in H_s^0, a solution of problem $R_{i_0,\ldots,i_s}$ minimizes the functional $(Au, u) - 2(f, u) = F(u)$, regarded on the class H_s^0.

Conversely, if for $u = u_0 \in H_s^0$ the functional $F(u)$ attains a minimum, then u_0 is a solution of the problem.

Suppose now that u_0 is a solution of problem $R_{i_0,\ldots,i_s}$ with nonhomogeneous boundary conditions in (1) and (5), and let v satisfy the same boundary conditions. Then $w_0 = u_0 - v \in H_s^0$ and is a solution of the equation $Aw = f_1$, where $f_1 = f - Av$. Consequently, w_0 minimizes the functional $F_1(w) = (Aw, w) - 2(f_1, w)$, regarded on H_s^0. Substituting $u - v$ in place of w in $F_1(w)$ and using the boundary conditions and (6), we get that $F_1(w) = \Phi(u) + \text{const}$, where

$$\Phi(u) = \int_\Omega \left(\sum a_{jk} \frac{\partial u}{\partial x_j} \frac{\partial u}{\partial x_k} + cu^2 \right) dx - 2(f, u) - 2 \sum_{m=s+1}^{N} \alpha_{j_m} q_{j_m}. \quad (7)$$

Obviously, w_0 minimizes $F_1(w)$ in H_s^0 if and only if $u_0 = w_0 + v$ minimizes the functional $\Phi(u)$ in the class $u \in H[i_0, \ldots, i_s; \alpha_{i_0}, \ldots, \alpha_{i_s}; q_{j_{s+1}}, \ldots, q_{j_N}]$.

Let us consider $\Phi(u)$ on the broader class $H[i_0, \ldots, i_s; \alpha_{i_0}, \ldots, \alpha_{i_s}]$. We show that if $\Phi(u)$ attains a minimum at a $u = u_0 \in H[i_0, \ldots, i_s; \alpha_{i_0}, \ldots, \alpha_{i_s}]$ then u_0 automatically satisfies the boundary condition (5), i.e., is a solution of problem $R_{i_0,\ldots,i_s}$.

Let $\eta \in H_s$, $\eta|_{\Sigma_{j_m}} = \eta_{j_m}$. Then for any t the function $u_0 + t\eta$ belongs to $H[i_0, \ldots, i_s; \alpha_{i_0,\ldots,}\alpha_{i_s}]$, and $\Phi(u_0 + t\eta) \geq \Phi(u_0)$. Hence, $\Phi'_t[u_0 + t\eta]|_{t=0} = 0$. We get that

$$\int_\Omega \left[\sum^n a_{jk} \frac{\partial \eta}{\partial x_j} \frac{\partial u_0}{\partial x_k} + c\eta u_0 - f\eta \right] dx - \sum_{m=s+1}^{N} \eta_{j_m} q_{j_m} = 0.$$

Using (6), we arrive at the equality

$$(Au_0 - f, \eta) + \sum_{m=s+1} \eta_{j_m} \left(\int_{\Sigma_{j_m}} \frac{\partial u}{\partial \nu} ds - q_{j_m} \right) = 0. \quad (8)$$

For any $\eta \in H_N$ $(H_s \supset H_N)$ formula (8) gives us that $(Au_0 - f, \eta) = 0$. From this it follows that $Au_0 - f = 0$, $x \in \Omega$. Condition (5) is a consequence of (8) for arbitrary η_{j_m}.

Accordingly, problem $R_{i_0,\ldots,i_s}$ reduces to the determination of a minimum for the functional (7) on the set $H[i_0, \ldots, i_s; \alpha_{i_0}, \ldots, \alpha_{i_s}]$.

§3. Construction of a system of coordinate functions in H_0

To find the minimum of (7) by the Ritz method it is required to construct a system of coordinate functions $\{\varphi_k\}$ satisfying the following conditions: 1) $\varphi_k \in H_0$; 2) $\varphi_1, \ldots, \varphi_m$ are linearly independent for any m; and 3) any function $u \in H_0$ can be approximated with any degree of accuracy by linear combinations of the functions $\{\varphi_k\}$. It is also assumed that the functions of the system $\{\varphi_k\}$ satisfy the condition $\varphi_k|_{\Sigma_j} = \delta_{kj}$. Then the mth approximation u_m of the solution u of problem $R_{i_0,\ldots,i_s}$ is found in the form

$$u_m = \sum_{k=1}^{s} \alpha_{i_k} \varphi_{i_k} + \sum_{k=s+1}^{N} c_{j_k} \varphi_{j_k} + \sum_{k=N+1}^{N+m} c_k \varphi_k.$$

Obviously, $u_m \in H[i_0,\ldots,i_s;\alpha_{i_0},\ldots,\alpha_{i_s}]$. The coefficients c_k are determined from the system of linear equations

$$\frac{\partial \Phi(u_m)}{\partial c_k} = 0, \qquad k = j_{s+1},\ldots,j_N, N+1,\ldots,N+m. \tag{10}$$

An investigation of the convergence of $\{u_m\}$ to the exact solution u for an arbitrary positive-definite operator is carried out in [1]. Therefore, we confine ourselves here to a remark on the construction of the system $\{\varphi_k\}$.

Suppose that we have constructed the functions $\varphi_1,\ldots,\varphi_N$. For an arbitrary function $u \in H_0$ ($u|_{\Sigma_i} = \alpha_i$, $i = 0,\ldots,N$, $\alpha_0 = 0$) the function $u - \sum_{i=1}^N \alpha_i\varphi_i = v$ vanishes on the boundary of Ω, i.e., belongs to H_N. If $\{\rho_k(x)\}$ is a system of coordinate functions in H_N, then v can be approximated with any accuracy by linear combinations of $\{\rho_k(x)\}$. Consequently, the system of functions $\varphi_1,\ldots,\varphi_N,\rho_1,\ldots,\rho_m,\ldots$ is a coordinate system in H_n. To construct a coordinate system in H_N it is known ([2], Chapter IV) to be sufficient to take any sequence of linearly independent smooth functions $\{\pi_k\}$ that is dense in $C(\overline{\Omega})$ (for example, a sequence of polynomials) and set $\rho_k = w\pi_k$, where $w \in H_N$ and $w > 0$ in Ω. Let $\varphi_{N+m} = w\pi_m$. Then the functions $\{\varphi_k\}$ form the desired sequence. The question thus rests in the construction of the functions $\varphi_1,\ldots,\varphi_N$ and w. For example, if we have found functions w_i, $i = 0,1,\ldots,N$, satisfying the condition

$$w_i|_{\Sigma_i} = 0, \quad w_i(x) > 0, \quad x \in \overline{\Omega} \setminus \Sigma_i,$$

then we can set

$$w = w_0 w_1 \cdots w_N, \qquad \varphi_r = \frac{w_0 \cdots w_{r-1} w_{r+1} \cdots w_N}{w_0 \cdots w_{r-1} w_{r+1} \cdots w_N + w_r}, \qquad r = 1,\ldots,N.$$

§4. The connection between solutions of problem R_0 and solutions of the Dirichlet problem

In solving problem R_0 for the homogeneous equation $Au = 0$ we arrive at the linear system (10) in the unknowns c_k, $k = 1,\ldots,m$; it can be written in the form

$$B\overline{c} = \sum_{k=1}^N q_k \overline{e}^{(k)}. \tag{11}$$

Here $\overline{c},\overline{e}^{(k)} \in R^m$, and $\overline{e}^{(k)}$ is the kth column of the $m \times m$ identity matrix. Introducing the column vectors φ with components $\varphi_1,\ldots,\varphi_m$, we write the mth approximation in the form $u_m = \sum_{k=1}^m c_k\varphi_k = (\overline{c},\overline{\varphi})$.

Suppose now that $v^{(k)}$ is a solution of the Dirichlet problem for the equation $Av = 0$ with boundary condition $v^{(k)}|_{\Sigma_j} = \delta_{kj}$. It minimizes the functional $\Psi(v) = (Av,v)$. The coefficients d_j of the approximate solution $v_m^{(k)} = \sum_{j=1}^m d_j\varphi_j = (\overline{d},\overline{\varphi})$ by the Ritz method satisfy the system

$$B\overline{d} = \delta_m^{(k)} e^{(k)}, \qquad d_k = 1,\ \delta_m^{(k)} = \Psi(v_m^{(k)}).$$

Hence,

$$u_m = \sum_{k=1}^{N} \frac{q_k}{\delta_m^{(k)}} v_m^{(k)}.$$

Estimates for $\|u - u_m\|$ can be obtained by using known estimates of the rate of convergence of $\{v_m^{(k)}\}$ to the solution $v^{(k)}$ of the Dirichlet problem.

BIBLIOGRAPHY

1. S. G. Mikhlin, *Variational methods in mathematical physics*, 2nd rev. ed., "Nauka", Moscow, 1970; English transl. of 1st ed., Pergamon Press, Oxford and Macmillan, New York, 1964.

2. L. V. Kantorovich and V. I. Krylov, *Approximate methods of higher analysis*, 3rd ed., GITTL, Moscow, 1950; English transl., Interscience, New York, and Noordhoff, Groningen, 1958.

Translated by H. H. McFADEN

Amer. Math. Soc. Transl.
(2) Vol. **133**, 1986

Some Properties of Interpolation Processes
in Spaces of Continuous Functions

UDC 517.51

N. B. TIKHOMIROV

Let T be a compact subset of a metric space M, C_T the space of all continuous functions on T, and $L(C_T, C_T)$ the space of all continuous linear mappings of C_T into itself. Consider a sequence $\{t_n\}$ of distinct points of T. Assume that there exists a sequence $\{x_n\}$ of elements of C_T such that the inequality $D_{k,n}(t_k) \neq 0$ holds for every n, where

$$
D_{k,n}(t) = \begin{vmatrix}
x_1(t) & x_2(t) & \cdots & x_n(t) \\
x_1(t_1) & x_2(t_1) & \cdots & x_n(t_1) \\
\hdotsfor{4} \\
x_1(t_{k-1}) & x_2(t_{k-1}) & \cdots & x_n(t_{k-1}) \\
x_1(t_{k+1}) & x_2(t_{k+1}) & \cdots & x_n(t_{k+1}) \\
\hdotsfor{4} \\
x_1(t_n) & x_2(t_n) & \cdots & x_n(t_n)
\end{vmatrix}
$$

This inequality holds automatically if $\{x_n\}$ is a Markov sequence, i.e., if every polynomial $p(t) = \sum_n^n \lambda_i x_i(t)$ has at most $n-1$ zeros on T, $n = 1, 2, \ldots$.

We define a mapping A_n of C_T into itself by

$$
A_n : x(t) \to A_n(x; t) = \sum_{k=1}^{n} x(t_k) l_{k,n}(t), \qquad l_{k,n}(t) = \frac{D_{k,n}(t)}{D_{k,n}(t_k)}.
$$

Obviously, $A_n \in L(C_T, C_T)$, and the value of the operator A_n on an element x is a polynomial of order at most n formed from the system $\{x_i\}_1^n$ and equal to $x(t)$ at the points $t_1, \ldots, t_n$. It is called an interpolation polynomial for x with respect to the system $\{x_n\}$ with the interpolation nodes $\{t_n\}$.

1980 *Mathematics Subject Classification* (1985 *Revision*). Primary 65D05, 41A30.

Translation of Application of Functional Analysis to Approximation Theory (V. N. Nikol'-skiĭ, editor), Kalinin. Gos. Univ., Kalinin, 1981, pp. 110–124; MR **83i**:41008.

©1986 American Mathematical Society
0065-9290/86 $1.00 + $.25 per page

In the case when $T = [a, b]$ and $x_n(t) = t^{n-1}$, $n = 1, 2, \ldots$, the polynomials $A_n(x; t)$ are the classical Lagrange interpolation polyomials.

Let F be the set of convergence of the interpolation process, i.e., the set of all points t in T such that

$$\lim_n A_n(x; t) = x(t) \qquad \forall x \in C_T.$$

Erdős [1] proved that the measure of the set of convergence for the classical Lagrange polynomials is equal to zero.

Define G_x to be the set of divergence of the interpolation process for $x(t)$. At each point t of this set the limit of the sequence $\{A_n(x; t)\}$ does not exist or is equal to infinity. In the classical case there are examples of continuous functions on an interval $[a, b]$ such that $mG_x = b - a$ for certain concrete collections of interpolation nodes. However, for an arbitrary matrix of interpolation nodes it has not been proved that there is a continuous function for which the measure of the set of divergence of the interpolation process is positive. Below we prove the existence of Markov systems such that the measure of the set of convergence and the measure of the set of divergence can be greater than zero.

The main result at the basis of the investigation is a theorem of Korovkin [2] establishing that for every Markov system of continuous functions on $[0, 1]$ there exists a perfect set F_0 of uniform convergence, i.e., such that $A_n(x; t) \rightrightarrows x(t)$, $t \in F_0$, $\forall x \in C[0, 1]$.

The process of constructing such a set and a sequence of interpolation nodes can be described as follows. Let $t_1 = 0$ and $t_1 = 1$, and call $[0, 1]$ the interval of first rank. From it we discard a certain interval (α_1, b_1), $0 < \alpha_1 < \beta_1 < 1$, and call $[0, \alpha_1]$ and $[\beta_1, 1]$ the intervals of second rank. Next, let $t_3 = \alpha_1$ and $t_4 = \beta_1$ and discard certain intervals (α_2, β_2) and (α_3, β_3), each from one of the intervals of second rank. Continued indefinitely, this process leads to the perfect set F_0 having the intervals (α_i, β_i) as its complementary intervals, and the endpoints of these intervals, numbered from the left pairwise in order of rank, are taken as the interpolation nodes.

Let

$$x_1(t), \ x_2(t), \ x_3(t), \ldots \tag{1}$$

be a Markov system of continuous functions on $[0, 1]$, $F_0 \subset [0, 1]$ a perfect set, and $\{t_n\}$ a sequence of interpolation nodes. Moreover, F_0 and $\{t_n\}$ are such that Korovkin's theorem holds, i.e.,

$$A_n(x; t) \rightrightarrows x(t), \qquad t \in F_0, \ \forall x \in C[0, 1].$$

We define a mapping f of $C[0, 1]$ into itself by setting

$$f \colon x(t) \to y(t) = \begin{cases} x(t) \text{ if } t \in F_0, \\ \text{linear on intervals complementary to } F_0. \end{cases}$$

Let $f(C[0, 1]) = C^*[0, 1]$ and let $\{y_n\}$ be the sequence of images of the system (1). The set $C^*[0, 1]$ is a subspace of $C[0, 1]$, and for each y in $C^*[0, 1]$ the sequence $\{A_n^*(y; t)\}$ of interpolation polynomials with respect to the system $\{y_n\}$ with nodes $\{t_n\}$ is defined.

THEOREM 1. *The sequence $\{A_n^*(y;t)\}$ converges uniformly on $[0,1]$ to $y(t)$ for every function y in $C^*[0,1]$.*

PROOF. Fix a $y \in C^*[0,1]$ and consider the sequence $\{A_n^*(y;t)\}$. If $t \in F_0$, then $A_n^*(y;t) = A_n(y;t)$, since for such t we have $y_n(t) = x_n(t)$, $n = 1,2,\ldots$. Therefore, for every $\varepsilon > 0$ and for $n > N_\varepsilon$,

$$|A_n^*(y;t) - y(t)| < \varepsilon.$$

The last inequality is true also when the point t belongs to some interval in the complement of F_0, since on each such interval the polynomial $A_n^*(y;t)$ and the function $y(t)$ are linear and their values at the endpoints of the interval differ by less than ε. This implies that $\|A_n^*(y) - y\| < \varepsilon$. The theorem is proved.

Note that $f(A_n(y)) = A_n^*(y) \in C^*[0,1]$; therefore, the space $C^*[0,1]$ is the linear closure of the system $\{y_n\}$.

THEOREM 2. *There exists a continuous function $x(t)$ such that the sequence $\{A_n(x;t)\}$ does not converge to $x(t)$ at any point of the complement CF_0.*

PROOF. Let

$$x(t) = \begin{cases} x_1(t) & \text{if } t \in F_0, \\ x_1(t) + (t - \alpha_i)(\beta_i - t) & \text{if } t \in (\alpha_i, \beta_i), \ i = 1,2,3\ldots. \end{cases}$$

Then $A_n(x;t) = A_n(x_1;t)$, since $x(t_k) = x_1(t_k)$, $k = 1,2,\ldots$. But it follows from the equalities, $A_n(x_1;t_k) = x_1(t_k)$ that $A_n(x_1;t) - x_1(t) \equiv 0$, since $x_1(t),\ldots,x_n(t)$ is a Tchebycheff system, and thus $A_n(x_1;t) \equiv x_1(t)$. Therefore,

$$\lim_n A_n(x;t) = x_1(t) \neq x(t),$$

if $t \in CF_0$.

LEMMA 1. *If system (1) is closed in $C[0,1]$, then for every $\varepsilon > 0$ and any function $x(t) \in C[0,1]$ there exists a polynomial $p(t) = \sum_{i=1}^m \lambda_i x_i(t)$ coinciding with $x(t)$ at the points $t_1,\ldots,t_n$ and such that $\|x - p\| < \varepsilon$.*

The proof is analogous to that in [3] for algebraic polynomials.

LEMMA 2. *If system (1) is closed in $C[0,1]$, then for every interval (α, β) in the complement of F_0 there exist a continuous function $y(t)$ and a sequence $\{n_k\}$ of positive integers, such that:*

$$1) \quad \lim_k A_{n_k}(y;t) = \infty \quad \text{if } t \in (\alpha, \beta);$$

$$2) \quad A_{n_k}(y;t) \rightrightarrows y(t) \quad \text{if } t \in [0,\alpha] \cup [\beta,1]$$

PROOF. Let $\sum_1^\infty a_i$ be a positive series convergent to 1. Define a continuous function $y_0(t)$ by

$$y_0(t) = \begin{cases} 0 & \text{if } t \notin (\alpha, \beta); \\ 1/a_1^2 & \text{if } t = (\alpha + \beta)/2; \\ \text{linear on the intervals } (\alpha, (\alpha + \beta)/2) \text{ and } ((\alpha + \beta)/2, \beta). \end{cases}$$

Since system (1) is closed in $C[0,1]$, there exists a polynomial $q(t) = \sum_{i=1}^{n_1} \lambda_i x_i(t)$ such that $\|y_0 - q\| < 1$. Take a continuous function $x(t)$ such that $x(t_k) = q(t_k)$, $k = 1, \ldots, n_1$, and $\|x\| < 1$. By Lemma 1, there exists a polynomial $p_1(t)$ coinciding with $x(t)$ at the points $t_1, \ldots, t_{n_1}$, and satisfying the inequality $\|x - p_1\| < 1$, and it can be assumed that $\|p_1\| < 1$. We have that

$$A_{n_1}(p_1; t) = A_{n_1}(x; t) = A_{n_1}(q; t) = q(t).$$

Therefore, $|A_{n_1}(p_1; t)| < 1$ if $t \notin (\alpha, \beta)$; but if $t \in (\alpha, \beta)$, then

$$A_{n_1}(p_1; t) = q(t) > y_0(t) - 1 = \frac{2}{(\beta - \alpha)a_1^2}\rho(t),$$

where $\rho(t) = \min\{t - \alpha, \beta - t\}$.

Note also that if the order of the polynomial p_1 is m, then $A_n(p_1; t) \equiv p_1(t)$ for $n \geq m$.

By arguing in a similar way we can get a sequence $\{p_i(t)\}$ of polynomials in system (1) and a sequence $\{n_i\}$ of positive integers such that the following conditions hold:

a) $|A_{n_i}(p_i; t)| < 1$ if $t \notin (\alpha, \beta)$;

b) $A_{n_i}(p_i; t) > \dfrac{2}{(\beta - \alpha)a_i^2}\rho(t) - 1$ if $t \in (\alpha, \beta)$;

c) $\|p_i\| < 1$;

d) $A_{n_{i+j}}(p_i; t) \equiv p_i(t)$.

We define a sequence $\{i_k\}$ of positive integers such that

$$\sum_{i=i_k}^{\infty} a_i < \frac{1}{(k-1)\|A_{n_{i_{k-1}}}\|}.$$

Then the series $\sum_{s=k+1}^{\infty} a_{i_s}$ has the property that

$$\sum_{s=k+1}^{\infty} a_{i_s} < \frac{1}{k\|A_{n_{i_k}}\|}.$$

Let $y(t) = \sum_{k=1}^{\infty} a_{i_k} p_{i_k}(t)$. It is clear that $y \in C[0,1]$ and $\|y\| < 1$. We have that

$$A_{n_{i_k}}(y; t) = A_{n_{i_k}}\left(\sum_{s=1}^{k-1} a_{i_s} P_{i_s}; t\right) + a_{i_k} A_{n_{i_k}}(p_{i_k}; t) + A_{n_{i_k}}\left(\sum_{s=k+1}^{\infty} a_{i_s} p_{i_s}; t\right).$$

By condition d),

$$A_{n_{i_k}}\left(\sum_{s=1}^{k-1} a_{i_s} p_{i_s}; t\right) = \sum_{s=1}^{k-1} a_{i_s} p_{i_s}(t);$$

and by c),

$$\left|A_{n_{i_k}}\left(\sum_{s=k+1}^{\infty} a_{i_s} p_{i_s}; t\right)\right| < \|A_{n_{i_k}}\| \sum_{s=k+1}^{\infty} a_{i_s} < \frac{1}{k}.$$

If $t \in (\alpha, \beta)$, inequality b) holds; since

$$\left|\sum_{s=1}^{k-1} a_{i_s} p_{i_s}(t)\right| < 1,$$

it follows that

$$A_{n_{i_k}}(y;t) > a_{i_k} A_{n_{i_k}}(p_{i_k};t) - 1 - \frac{1}{k} > \frac{2}{(\beta - \alpha)a_{i_k}}\rho(t) - 3.$$

Therefore, for such t

$$\lim_k A_{n_{i_k}}(y;t) = \infty.$$

If $t \notin (\alpha, \beta)$, then condition a) holds, and, consequently,

$$\left| A_{n_{i_k}}(y;t) - \sum_{s=1}^{k-1} a_{i_s} p_{i_s}(t) \right| < a_{i+k} + \frac{1}{k}.$$

In this case

$$A_{n_{i_k}}(y;t) \rightrightarrows y(t), \qquad t \in [0,\alpha] \cup [\beta, 1],$$

which is what we were required to prove.

For what follows note that the last inequality implies that $|A_{n_{i_k}}(y;t)| < 3$ if $t \notin (\alpha, \beta)$.

THEOREM 3. *If system* (1) *is closed in* $C[0,1]$, *then there exist a continuous function* $y(t)$ *and a sequence* $\{n_k\}$ *of positive integers such that at each point* $t \in CF_0$

$$\lim_k A_{n_k}(y;t) = \infty.$$

PROOF. Consider the sequence $\{(\alpha_i, \beta_i)\}$ of open intervals forming the complement of F_0. According to Lemma 2, there exist a function $y(t) \in C[0,1]$ and a sequence of positive integers $n_1^{(1)} < n_2^{(1)} < \cdots$, such that

$$A_{n_k^{(1)}}(y_1;t) \to \infty \quad \text{if } t \in (\alpha_1, \beta_1);$$

$$\|y_1\| < 1;$$

$$|A_{n_k^{(1)}}(y_1,t)| < 3 \quad \text{if } t \notin (\alpha_1, \beta_1).$$

We take the interval (α_2, β_2) and find a function $y_2(t) \in C[0,1]$ and a corresponding sequence of indices $n_1^{(2)} < n_2^{(2)} < \cdots$ such that

$$A_{n_k^{(2)}}(y_2;t) \to \infty \quad \text{if } t \in (\alpha_2, \beta_2);$$

$$\|y_2\| < 1;$$

$$|A_{n_k^{(2)}}(y_2;t)| < 3 \quad \text{if } t \notin (\alpha_2, \beta_2).$$

Moreover, it can be assumed that the sequence $\{n_k^{(2)}\}$ is a subsequence of $\{n_k^{(1)}\}$.

Continuing this process, we get a sequence $\{y_i(t)\}$ of continuous functions and a matrix of indices

$$
\begin{array}{ccccc}
n_1^{(1)} & n_2^{(1)} & \cdots & n_k^{(1)} & \cdots \\
n_1^{(2)} & n_2^{(2)} & \cdots & n_k^{(2)} & \cdots \\
\cdots\cdots\cdots\cdots\cdots\cdots\cdots\cdots\cdots \\
n_1^{(i)} & n_2^{(i)} & \cdots & n_k^{(i)} & \cdots \\
\cdots\cdots\cdots\cdots\cdots\cdots\cdots\cdots\cdots
\end{array}
$$

such that

$$A_{n_k^{(i)}}(y_i;t) \to \infty \quad \text{if } t \in (\alpha_i;\beta_i);$$

$$\|y_i\| < 1;$$

$$|A_{n_k^{(i)}}(y_i;t)| < 3 \quad \text{if } t \notin (\alpha_i;\beta_i).$$

For every $i > 1$ the sequence $\{n_k^{(i)}\}$ is a subsequence of $\{n_k^{(i-1)}\}$.

Let $\sum_1^\infty b_i$ be a positive convergent series satisfying the conditions

$$\sum_{i=k+1}^\infty b_i < \frac{1}{k\|A_{n_k^{(k)}}\|}, \qquad k = 1,2,3,\ldots,$$

and let $y(t) = \sum_1^\infty b_i y_i(t)$. We show that at each point $t \in CF_0$

$$\lim_k A_{n_k^{(k)}}(y;t) = \infty.$$

Let $t \in (\alpha_i,\beta_i)$. Then for $k > i$

$$A_{n_k^{(k)}}(y;t) = \sum_{\substack{s=1 \\ s\neq i}}^{k-1} b_s A_{n_k^{(k)}}(y_s;t) + b_i A_{n_k^{(k)}}(y_i;t)$$

$$+ A_{n_k^{(k)}}\left(\sum_{s=k+1}^\infty b_s y_s;t\right).$$

Since $k > i$, $\{A_{n_k^{(k)}}\}$ is a subsequence of $\{A_{n_k^{(i)}}\}$; therefore,

$$\lim_k A_{n_k^{(k)}}(y_i;t) = \infty.$$

On the other hand,

$$\left|A_{n_k^{(k)}}\left(\sum_{s=k+1}^\infty b_s y_s;t\right)\right| < \|A_{n_k^{(k)}}\| \sum_{s=k+1}^\infty b_s < \frac{1}{k}$$

and

$$\left|\sum_{\substack{s=1 \\ s\neq i}}^{k-1} b_s A_{n_k^{(k)}}(y_s;t)\right\| < 3\sum_{s=1}^{k-1} b_s < 3.$$

Therefore,

$$A_{n_k^{(k)}}(y;t) > b_i A_{n_k^{(k)}}(y_i;t) - 4,$$

and the theorem follows from this inequality.

COROLLARY. *If system* (1) *is closed in* $C[0,1]$, *then the sequence* $\{\|A_n\|\}$ *is not bounded.*

Suppose now that T is a compact subset of a metric space M. It is known ([4], Chapter V, §1, Theorem 6) that T is a continuous image of the Cantor perfect set P_0. Assume that T and P_0 are homeomorphic; then T and F_0 are homeomorphic, since P_0 and F_0 are homeomorphic. In this case, according to the Banach-Stone theorem ([5], Chapter V, §3, Theorem 2) the spaces C_T and C_{F_0} are linearly isometric. Moreover, the formula $f: x \mapsto x \circ \varphi^{-1}$ determines an isometry of C_{F_0} onto C_T, where φ is a homeomorphism of T onto F_0.

THEOREM 4. *If the compact set T is homeomorphic to the Cantor perfect set P_0, then there exist a sequence $\{u_n\}$ $(u_n \in T)$ of interpolation nodes and a Markov system $\{y_n\}$ $(y_n \in C_T)$ such that the sequence $\{A_n^*(y)\}$ of interpolation polynomials of the system $\{y_n\}$ with the nodes $\{u_n\}$ converges in the C_T-norm to y for every function $y(u) \in C_T$.*

PROOF. It follows from the condition of the theorem that T and F_0 are homeomorphic.

Let φ be a homeomorphism of T onto F_0, f the isometry it generates from C_{F_0} onto C_T, $\{x_n\}$ a Markov system in C_{F_0}, and $\{t_n\}$ a sequence of interpolation nodes such that

$$A_n(x; t) \rightrightarrows x(t), \qquad t \in F_0, \ \forall x \in C_{F_0}.$$

Let $u_n = \varphi^{-1}(t_n)$ and $y_n = f(x_n)$, $n = 1, 2, \ldots$. Obviously, $\{y_n\}$ is a Markov system, since the polynomials $\sum_1^n \lambda_i y_i(u)$ and $\sum_1^n \lambda_i x_i(t)$ have the same number of zeros.

Let $y \in C_T$ and $x = f^{-1}(y)$. It is not hard to see that $f[A_n(x)] = A_n^*(y)$. Indeed,

$$f[A_n(x)] = f[A_n(x; t)] = f\left[\sum_{k=1}^{n} x(t_k) l_{k,n}(t)\right]$$

$$= \sum_{k=1}^{n} (x \circ \varphi^{-1})(u_k) f[l_{k,n}(\varphi(u))]$$

$$= \sum_{k=1}^{n} y(u_k) l_{k,n}^*(u) = A_n^*(y; u).$$

Therefore, since f is an isometric mapping,

$$\|A_n^*(y) - y\| = \|A_n(x) - x\| \to 0 \quad \text{as } n \to \infty.$$

This implies the statement of the theorem.

Let F be a set of points in $[0, 1]$. By using the theorems given in [6] (§§36.I, 35.II) it is not hard to show that F is homeomorphic to P_0 if and only if it is a perfect nowhere dense set. Moreover, a homeomorphism of F onto P_0 can be extended to a homemorphism of $[0, 1]$ onto itself if the endpoints belong to F.

THEOREM 5. *For every number δ with $0 \leq \delta < 1$ there exist a Markov system of continuous functions on $[0, 1]$, a perfect set $F \subset [0, 1]$ of measure δ, and a matrix of interpolation nodes on it such that the following conditions hold:*

1) The sequence of interpolation polynomials of the given system converges uniformly on F for every function $y \in C[0, 1]$.

2) There exists a function $y_0 \in C[0, 1]$ such that the sequence of interpolation polynomials does not converge to it at any point of CF.

3) There exists a function $y_1 \in C[0, 1]$ such that the sequence of interpolation polynomials diverges at each point of CF.

PROOF. We construct on $[0, 1]$ a perfect nowhere dense set F of measure δ.

Let F_0 be the set of uniform convergence of the interpolation process with respect to the system of powers $1, t, t^2, \ldots$. According to the foregoing there exists a homeomorphism φ of $[0, 1]$ onto itself which maps F onto F_0, and the mapping f of $C[0, 1]$ onto itself defined by $f : x \mapsto x \circ \varphi^{-1}$ is a linear isometry. Moreover, $f(t^n) = [\varphi^{-1}(t)]^n$. The system $\{[\varphi^{-1}(t)]^n\}$ is a Markov system and is closed in $C[0, 1]$.

By Theorem 4, the interpolation polynomials of this system converge uniformly on F for every continuous function on $[0, 1]$.

It follows from Theorem 2 that there is a continuous function such that the sequence of interpolation polynomials does not coverge to it at any point of CF.

On the basis of Theorem 3 there exists a continuous function whose sequence of interpolation polynomials diverges at each point of CF. The theorem is proved.

REMARKS. 1. For the case of algebraic interpolation polynomials there exists a perfect set F of measure zero such that for any matrix of interpolation nodes there is a continuous function whose sequence of interpolation polynomials diverges on a dense subset of F. An example is given in [7].

2. It follows from [8] that all our results can be formulated also in the case when the interpolation nodes can be taken to be points such that all the nodes in the nth row of the matrix differ from the nodes in the remaining rows.

3. Theorems 5 and 1 give us a theorem noted earlier by us in [9] on uniform convergence of interpolation processes in subspaces of $C[0, 1]$.

BIBLIOGRAPHY

1. P. Erdős, *Problems and results on the theory of interpolation.* I, Acta Math. Acad. Sci. Hungar. **9** (1958), 381–388.
2. P. P. Korovkin, *On the closedness of the system of Tchebycheff functions*, Dokl. Akad. Nauk SSSR **78** (1951), 853–855. (Russian)
3. I. P. Natanson, *Constructive function theory*, GITTL, Moscow, 1949; English transl., Vols. I, II, III, Ungar, New York, 1964, 1965.
4. L. A. Lyusternik and V. I. Sobolev, *Elements of functional analysis*, 2nd rev. ed., "Nauka", Moscow, 1965; English transl. of 1st. ed., Ungar, New York, 1961.
5. Mahlon M. Day, *Normed linear spaces*, Springer-Verlag, 1958.
6. K. Kuratowski, *Topology*, Vol. I, 5th ed., PWN, Warsaw, and Academic Press, New York, 1966.
7. N. B. Tikhomirov, *Special questions in the theory of interpolation of functions*, Lecture notes, Kalinin. Gos. Univ., Kalinin, 1973, pp. 73–78. (Russian)
8. V. V. Lokot', *On an interpolation process of Lagrange type*, Kalinin. Gos. Ped. Inst. Uchen. Zap. **39** (1964), 10–17.
9. N. B. Tikhomirov, *On the approximation of continuous functions by interpolation polynomials*, Theory of Approximation of Functions (Proc. Internat. Conf., Kaluga, 1975), "Nauka", Moscow, 1977, pp. 365–366. (Russian)

Translated by H. H. MCFADEN

Amer. Math. Soc. Transl.
(2) Vol. **133**, 1986

On Multipliers of Double Fourier Series

S. BARON

One of the difficult problems of the theory of trigonometric series is that of determining from the coefficients of a trigonometric series whether that series is the Fourier series of a function from a prescribed class of functions. Many mathematicians have studied this problem (cf., for example, [24], p. 82). In 1923 Kolmogorov proved that a cosine series $\sum a_n \cos nx$ is the Fourier series of a Lebesgue-integrable function if its coefficients a_n tend to zero and (a_n) is a quasiconvex sequence (cf. [1], p. 652 (English Vol. II, p. 202)). In 1933–34 Moore and Cesari generalized Kolmogorov's result to more general hypotheses with respect to (a_n). In 1960 Goes ([25], pp. 136–137) gave a new proof of the Moore-Cesari theorem, applying theorems on summability factors for the Cesàro method C^α of order $\alpha > 0$ (cf., for example, [5], Theorems 22.3 and 22.2 with $\beta = \alpha$), and found conditions under which a cosine series is a Fourier-Stieltjes series ([10], p. 11). Tönnov ([24], Corollaries 3.1 and 3.4), generalizing Goes' result, showed that $\sum a_n \cos nx$ is the Fourier series of an integrable function (resp. a Fourier-Stieltjes series) if the a_n are summability factors of type (T_0, T_1) (resp. (T, T_1)), where T and T_1 are arbitrary regular summation methods and T sums uniformly the Fourier series of all 2π-periodic continuous functions. Further, knowledge of effective conditions for determining a class of functions from the coefficients of Fourier cosine series makes it easy to find effective conditions for multipliers of Fourier series (cf. [24], p. 84, or [5], p. 248). This also has been studied by many mathematicians, starting in 1923 (cf. for example [10], p. 378).

In the present article we solve analogous problems for double trigonometric series. We first find conditions for determining the class of a function of two variables from the coefficients of its double Fourier series, and then, applying

1980 *Mathematics Subject Classification* (1985 *Revision*). Primary 42B15.

Translation of Tartu Riikl. Ül. Toimetised No. 504(1981), 116–125; MR **84j**:42006.

©1986 American Mathematical Society
0065-9290/86 $1.00 + $.25 per page

these conditions and certain theorems of Skvortsova [17]–[19], we obtain conditions for multipliers of double Fourier series. These conditions are expressed in terms of summation multipliers of double series. We recall that the numbers ε_{mn} are called *summation multipliers of type* (S_0, T_r) (or of type (S_r, T_r)) if for any S-bounded (resp. S-summable) double series[1] $\sum u_{mn}$ the double series $\sum \varepsilon_{mn} u_{mn}$ is completely T-summable.

§1.   Lemmas on T-complementary spaces

Let f and g be real-valued functions of two variables defined almost everywhere on $\mathbf{R}^2$, periodic with period 2π in each variable, and Lebesgue-integrable on the square $Q = [-\pi, \pi]^2$. Let f and g have the double Fourier series expansions

$$f^\circ = \sum A_{kl}, \tag{1}$$

$$g^\circ = \sum \mathbf{A}_{kl}, \tag{2}$$

respectively, where

$$A_{kl}(s, t) = \lambda_{kl}(a_{kl} \cos ks \cos lt + b_{kl} \sin ks \cos lt$$

$$+ c_{kl} \cos ks \sin lt + d_{kl} \sin ks \sin lt),$$

$$\mathbf{A}_{kl}(s, t) = \lambda_{kl}(\alpha_{kl} \cos ks \cos lt + \beta_{kl} \sin ks \cos lt$$

$$+ \gamma_{kl} \cos ks \sin lt + \delta_{kl} \sin ks \sin lt),$$

and $\lambda_{00} = \frac{1}{4}$, $\lambda_{kl} = \frac{1}{2}$ if $kl = 0$ with $k + l > 0$, and $\lambda_{kl} = 1$ for $k, l > 0$.

We shall identify functions which are equal almost everywhere on Q.

We shall use the same symbol to denote both a set of two-variable functions f and the set of their double Fourier series f°. For example we shall denote by C the space of all continuous functions f on Q and also the space of all double Fourier series (1) of these functions.

Let X be some space of double series (1). Let $T = (\tau_{mnkl})$ be a triangular summation method.[2] Following Goes and Tönnov (cf. [5], p. 241) we shall call the space of all trigonometric series (2) for which the double numerical series

$$\sum \lambda_{kl}(\alpha_{kl} a_{kl} + \beta_{kl} b_{kl} + \gamma_{kl} c_{kl} + \delta_{kl} d_{kl}) \tag{3}$$

is completely T-summable for every f° in X T-*complementary to* X, and we denote this space by $(X \to T)$. If $a_{kl} = 0$ for all f° in X, we shall also suppose $\alpha_{kl} = 0$ for all g° in $(X \to T)$. We make analogous conventions with regard to the coefficients b_{kl}, c_{kl}, and d_{kl}. Thus, if X consists entirely of cosine-cosine series

$$c^\circ(s, t) = \sum \lambda_{kl} a_{kl} \cos ks \cos lt, \tag{4}$$

then $(X \to T)$ is the same, i.e. consists of cosine-cosine series

$$k^\circ(s, t) = \sum \lambda_{kl} \alpha_{kl} \cos ks \cos lt. \tag{5}$$

---

[1] If limits of summation are not indicated, they are 0 and ∞ on each index.

[2] We shall give summation methods by means of transformation matrices of a double series into a double sequence. Free indices assume all values $0, 1, \ldots$.

Let $S = (\sigma_{mnkl})$ be some triangular method of summing double series. Let X consist of all cosine-cosine series (4) for which $\sum \lambda_{kl}a_{kl} \in S_0$ (resp. $\sum \lambda_{kl}a_{kl}$ belongs to S_r).[3] Then (by definition) the T-complementary space $(X \to T)$ consists of all double series (5) for which the double series $\sum \alpha_{kl}\lambda_{kl}a_{kl}$ is completely T-summable. The latter means that the numbers α_{kl} are summation multipliers of type (S_0, T_r) (resp. (S_r, T_r)).

Thus we have proved

LEMMA 1. *If the set X consists of all series (4) with S-bounded (resp. completely S-summable) coefficients, then the T-complementary space $(X \to T)$ consists of those series (5) whose coefficients α_{kl} are summation multipliers of type (S_0, T_r) (resp. (S_r, T_r)).*

Let L be the space of all functions f which are Lebesgue-integrable on Q and let M be the space of all functions f in L which are bounded almost everywhere on Q. Let us see which double series (1) the T-complementary space $(M \to T)$ consists of.

Denote the S- and T-means of the series (1) by $\sigma_{mn}f$ and $\tau_{mn}f$, i.e.

$$\sigma_{mn}f = \sum_{k,l \leq m,n} \sigma_{mnkl}A_{kl}, \qquad \tau_{mn}f = \sum_{k,l \leq m,n} \tau_{mnkl}A_{kl},$$

and the T-mean of the series (3) by h_{mn}, i.e.

$$h_{mn} = \sum_{k,l \leq m,n} \tau_{mnkl}\lambda_{kl}(\alpha_{kl}a_{kl} + \beta_{kl}b_{kl} + \gamma_{kl}c_{kl} + \delta_{kl}d_{kl}).$$

By the definition of a T-complementary space a double series g° belongs to $(M \to T)$ if and only if the numerical double series (3) is completely T-summable, i.e. the numerical double sequence (h_{mn}) converges completely for every $f \in M$. To each f in M we associate a function g in $(M \to T)$ using (2) and (3), and define a functional $\phi\colon L \to \mathbf{R}$ by the formula

$$\varphi(\tau_{mn}g) = h_{mn}.$$

Now, taking account of the general form of a continuous linear functional on L (cf. [13], p. 257 (English p. 191)) we see that the convergence of the double sequence (h_{mn}) entails the weak convergence of $(\tau_{mn}g)$ in L. For, expressing the Fourier coefficients a_{kl}, b_{kl}, c_{kl}, and d_{kl} of the function f by integrals (cf., for example, [23], Russian p. 222, German p. 165, English 176, or [9], p. 122), we obtain

$$h_{mn} = \frac{1}{\pi}^2 \iint\limits_Q f(x,y)(\tau_{mn}g)(x,y)\,dx\,dy. \tag{6}$$

We shall show that $(\tau_{mn}g)$ is weakly completely convergent to g. To do this it suffices to show that the double Fourier series of the weak limit coincides

<hr>

[3]S_0 (resp. S_r) denotes the set of all S-bounded (resp. completely S-summable) double series.

with g°. Indeed, setting first $f(x,y) = \cos\mu x \cdot \cos\nu y$ in (6), we obtain $h_{mn} = \tau_{mn\mu\nu}\lambda_{\mu\nu}\alpha_{\mu\nu}$; then, setting $f(x,y) = \sin\mu x \cdot \cos\nu y$, we find $h_{mn} = \tau_{mn\mu\nu}\lambda_{\mu\nu}\beta_{\mu\nu}$ and so forth, whence, if T satisfies the condition(^{4})

$$r\text{-}\lim_{m,n} \tau_{mnkl} = 1, \tag{7}$$

the desired result follows. Thus we have proved

LEMMA 2. *If T satisfies condition (7), then the T-complementary space $(M \to T)$ consists of those g° in L such that $(\tau_{mn}g)$ is weakly completely convergent to g in L.*

2.   Finding conditions for determining a class of functions

Let K_{mn} be the kernel of the method S, i.e. (cf. [5], p. 242)

$$K_{mn}(u,v) = \sum_{k,l \leq m,n} (\Delta_{kl}\sigma_{mnkl})D_k(u)D_l(v)$$

where D_k is the Dirichlet kernel. Let L_{mn} be the Lebesgue constants of the method S, i.e.

$$L_{mn} = \frac{1}{\pi^2} \iint\limits_{Q} |K_{mn}(u,v)|\,du\,dv.$$

Suppose the method S satisfies the condition

$$L_{mn} = O(1). \tag{8}$$

Then, taking account of the fact that

$$(\sigma_{mn}f)(s,t) = \frac{1}{\pi^2} \iint\limits_{Q} f(s+u, t+v)K_{mn}(u,v)\,du\,dv \tag{9}$$

(cf. [11], p. 303), for all f in M, we obtain

$$|(\sigma_{mn}f)(s,t)| \leq L_{mn}\|f\|_M, \tag{10}$$

where

$$\|f\|_M = \operatorname*{ess\,sup}_{Q} |f(s,t)|.$$

Hence, in view of (8) it follows that everywhere on Q

$$(\sigma_{mn}f)(s,t) = O(1),$$

i.e. all f° in M are S-bounded everywhere on Q and, consequently, $\sum \lambda_{kl}a_{kl} \in S_0$, for $f^\circ(0,0) = \sum \lambda_{kl}a_{kl}$. Hence using the notation

$$X_0 = \{c^\circ \colon \sum \lambda_{kl}a_{kl} \in S_0\},$$

we obtain that $M_c \subset X_0$, where M_c is the set of all even functions which are bounded almost everywhere on Q (cf. [9], p. 149). Then by the definition of

---

(^{4})The notation $r\text{-}\lim_{m,n} x_{mn} = a$ means that (x_{mn}) converges completely to a, i.e. $\lim_{n\to\infty} x_{mn}$ exists, $\lim_{m\to\infty} x_{mn}$ exists, and $\lim_{m,n\to\infty} x_{mn} = a$.

a T-complementary space we have $(X_0 \to T) \subset (M_c \to T) \subset (M \to T)$. By
Lemma 1 the series k° belongs to $(M \to T)$ if the numbers α_{mn} are summation
multipliers of type (S_0, T_r), i.e. if $\sum \lambda_{kl}(\alpha_{kl}a_{kl} + \beta_{kl}0 + \gamma_{kl}0 + \delta_{kl}0)$ is in T_r
for every f° in M. By Lemma 2 we have $(M \to T) \subset L$. Thus if the α_{kl} are
summation multipliers of type (S_0, T_r), then $k^\circ \in L$. And so we have proved

THEOREM 1. *Suppose the triangular methods S and T satisfy respectively
conditions* (8) *and* (7). *If the numbers a_{mn} are summation multipliers of type
(S_0, T_r), then the cosine-cosine series* (4) *is a Fourier series from L.*

Let V be the space of all functions of two variables having bounded variation
in the sense of Vitali on Q (cf. [9], p. 220). We shall denote by dV the space
of all double Fourier-Stieltjes series of functions in V, i.e. double series (1) for
which the Fourier-Stieltjes coefficients a_{kl} of a function F in V are defined by

$$a_{kl} = \frac{1}{\pi^2} \iint\limits_{Q} \cos kx \cos ly \, dF(x, y)$$

(cf. [21], p. 246) and the coefficients b_{kl}, c_{kl}, and d_{kl} are calculated from analogous
formulas.

We shall prove

THEOREM 2. *Suppose the triangular methods S and T satisfy respectively
conditions* (8) *and* (7), *and that*

$$r\text{-}\lim_{m,n} \sigma_{mnkl} = 1. \tag{11}$$

*If the numbers a_{mn} are summation multipliers of type (S_r, T_r) then the cosine-
cosine series* (4) *is a Fourier-Stieltjes series.*

PROOF. We first prove that in view of (11) and (8) all f° in C are uniformly
completely S-summable on Q. Indeed, from (10) it follows that for the operators
$\sigma_{mn}: C \to C$ defined by (9)

$$\|\sigma_{mn}\| \leq L_{mn}. \tag{12}$$

Further, by the Weierstrass approximation theorem (cf. [8], p. 110, or [22], p. 14
(English p. 5)) the set of double trigonometric polynomials is dense in C and,
consequently, the set

$$E = \{1/2, \cos \mu s, \sin \mu s\} \times \{1/2, \cos \nu t, \sin \nu t\}, \quad \mu, \nu \geq 1,$$

is fundamental in C. But, in view of (11) for each f in E the double sequence
defined in (9) is uniformly completely convergent on Q, since for each f in E
we have $\|f - \sigma_{mn}f\| = |1 - \sigma_{mn\mu\nu}| \|f\|$. In view of (12) and (8) we obtain the
desired result from the generalization of the Hahn-Banach-Steinhaus theorem
given by Kull' (cf. [14], p. 10, Theorem III), i.e. for each f in C

$$r\text{-}\lim_{m,n} \|f - \sigma_{mn}f\| = 0.$$

And so $C \subset S_r$, whence for all f° in C we have $\sum \lambda_{kl} a_{kl} \in S_r$. Hence, using the notation

$$X_r = \{ c^\circ \colon \sum \lambda_{kl} a_{kl} \in S_r \},$$

we obtain that $C_c \subset X_r$, where C_c is the set of all even functions which are continuous on Q. Then by the definition of a T-complementary space we have $(X_r \to T) \subset (C_c \to T) \subset (C \to T)$, whence by Lemma 1 we conclude that $k^\circ \in (C \to T)$ if the numbers α_{mn} are summation multipliers of type (S_r, T_r), i.e. if $\sum \alpha_{kl} \lambda_{kl} a_{kl} \in T_r$ for every $f^\circ \in C$. This is the case when the double sequence (h_{mn}) defined by (6) is completely convergent for every f in C. Hence $k^\circ \in dV$. To prove this we define a sequence of linear functionals $\psi_{mn} \colon C \to \mathbf{R}$ which is pointwise completely convergent on C, by means of the formula $\psi_{mn} f = h_{mn}$. From the Riesz-Markov theorem (cf. [16], English p. 129, or [20], p. 81 (English p. 70)) on the general form of a continuous linear functional on C it follows that all ψ_{mn} are continuous on C. Then the limiting functional ψ is also linear and continuous on C (cf. [14], p. 12, Theorem IV). By the Riesz-Markov theorem the functional ψ is defined by

$$\psi f = \frac{1}{\pi^2} \iint\limits_Q f(s,t) \, dG(s,t)$$

with $G \in V$. Hence for $f(x,y) = \cos \mu x \cdot \cos \nu y$, taking account of (6) and (7), we deduce that

$$\psi f = \lim_{m,n} \tau_{mn\mu\nu} \lambda_{\mu\nu} \alpha_{\mu\nu} = \lambda_{\mu\nu} \alpha_{\mu\nu}.$$

Consequently (α_{kl}) is the sequence of coefficients of a Fourier-Stieltjes series, i.e. $k^\circ \in dV$. The theorem is proved.

§3. Finding conditions for multipliers

A double sequence (a_{mn}) is called a *multiplier of class* (X, Y) if the double series $\sum a_{kl} A_{kl}$ is the double Fourier series of a function from Y whenever $g^\circ \in X$.

We shall prove

THEOREM 3. *Suppose the triangular methods S and T satisfy respectively conditions (8) and (7). If the numbers a_{mn} are summation multipliers of type (S_0, T_r), then (a_{mn}) is a multiplier of class (dV, L).*

PROOF. Skvortsova (cf. [17], Theorem 4) has proved that (a_{mn}) is a multiplier of class (dV, L) if and only if $c^\circ \in L$. Applying Theorem 1, we obtain the desired result.

Let ϕ be a convex continuous function of one variable $u \geq 0$ such that $0 \leq \phi(u) \uparrow$, $\phi(2u) = O(1)\phi(u)$, and $u^{-1}\phi(u) \to \infty$ as $u \to \infty$. We shall denote by L_ϕ the class of functions f for which

$$\iint\limits_Q \phi(|f(x,y)|) dx \, dy < \infty$$

(cf. [19], p. 19). In particular, for $\phi(u) = u^p$ with $p > 1$ we write L^p instead of L_ϕ.

We shall denote by R the class of Riemann-integrable functions of two variables on Q.

THEOREM 4. *Suppose the methods S and T are triangular, S satisfies conditions (8) and (11), and T satisfies (7). If the numbers a_{mn} are summation multipliers of type (S_r, T_r), then (a_{mn}) is a multiplier of classes (M, M), (R, R), (C, C), (dV, dV), (R, M), (C, M), (C, R), (L_ϕ, L_ϕ), and (L^p, L^p).*

PROOF. Skvortsova (cf. [17], Theorem 2, [18], Theorem 7, and [19], Theorem 12) has proved that (a_{mn}) is a multiplier of the first seven classes of Theorem 4 if and only if $c^\circ \in dV$ and a multiplier of the class (L_ϕ, L_ϕ) if $c^\circ \in dV$. Applying Theorem 2, we now obtain the desired result.

Theorems 3 and 4 reduce the problem of finding conditions for the multipliers of double Fourier series under consideration to the problem of multipliers of summation types (S_0, T_r) and (S_r, T_r), which has been solved for Cesàro summation ([2], Theorems 3 and 1) and for weighted Riesz means ([3], Theorems 6 and 4), as well as for arbitrary methods S and T satisfying certain conditions (cf. [7], Theorem 1, and [6], Theorems 3 and 1), or when T is an E convergence method (cf. [12], Theorem 14, and [4], Theorems 5 and 3). Thus we obtain for the multipliers effective conditions imposed directly on them.

We shall apply Theorems 3 and 4 in the case when S and T are respectively the Cesàro methods $C^{\alpha,\beta}$ with $\alpha, \beta > 0$ and $C^{\gamma,\delta}$ with $\gamma, \delta \geq 0$. Conditions (11) and (7) are satisfied for them. That condition (8) holds follows from Nikol'skiĭ's theorem (cf. [1], p. 476 (English Vol. II, p. 5), or [15], pp. 265 and 267) or can be obtained by direct computation (cf. [10], p. 94) in view of the fact that Cesàro methods are factorable. Applying now Theorems 3 and 1 of [2], we obtain

COROLLARY 1. *If for $\alpha, \beta > 0$ and $0 \leq \gamma, \delta \leq \alpha, \beta$ the conditions $a_{mn} = O(1)$ and*

$$\sum_{m,n} (m+1)^\alpha (n+1)^\beta |\Delta_{mn}^{\alpha+1\ \beta+1} a_{mn}| < \infty, \tag{13}$$

$$\sum_{m} (m+1)^\alpha |\Delta_m^{\alpha+1} a_{mn}| = o[(n+1)^{\beta-\delta}], \tag{14}$$

$$\sum_{n} (n+1)^\beta |\Delta_n^{\beta+1} a_{mn}| = o[(m+1)^{\gamma-\alpha}], \tag{15}$$

$$\lim_{m,n}[(m+1)^{\alpha-\gamma}(n+1)^{\beta-\delta} a_{mn}] = 0,$$

are satisfied, then (a_{mn}) is a multiplier of class (dV, L), while if condition (13) and the condition

$$a_{mn} = O[(m+1)^{\gamma-\alpha}(n+1)^{\delta-\beta}],$$

are fulfilled, and o is replaced by O in (14) and (15), then (a_{mn}) is a multiplier of all classes of Theorem 4.

We shall also apply Theorems 3 and 4 in the case when S is the factorable Voronoï-Nörlund summation method with $\sigma_{mnkl} = P_{m-k,n-l}/P_{mn}$, where $P_{mn} = P'_m P''_n > 0$ and $T = E$. Then defining the numbers d_{mn} by the double power series

$$\sum d_{mn} x^m y^n = \left(\sum P_{mn} x^m y^n \right)^{-1},$$

we obtain

COROLLARY 2. *If $P'_m - P'_{m-1}, P''_n - P''_{n-1} \downarrow 0$ and $P'_m, P''_n \to \infty$ while*

$$\sum_{k,l \leq m,n} (k+1)^{-1}(l+1)^{-1} P_{kl} = O(P_{mn})$$

and $\Sigma(m+1)(n+1)|d_{mn}| < \infty$, then (a_{mn}) is a multiplier of class (dV, L) if the following conditions hold:

$$r\text{-}\lim_{m,n}(P_{mn} a_{mn}) = 0,$$

$$\sum_{m,n} P_{mn} \left| \sum_{k,l \geq m,n} d_{k-m,l-n} a_{kl} \right| < \infty, \tag{16}$$

$$\sum_{m} P_{mn} \left| \sum_{k \geq m} d_{k-m,0} a_{kn} \right| = o(1), \tag{17}$$

$$\sum_{n} P_{mn} \left| \sum_{l \geq n} d_{0,l-n} a_{ml} \right| = o(1), \tag{18}$$

while if condition (16) holds and $P_{mn} a_{mn} = O(1)$ and o is replaced by O in (17) and (18), then (a_{mn}) is a multiplier of all the classes of Theorem 4.

PROOF. Conditions (11) and (7) are satisfied, for the factors of the method S are regular (cf. [5], p. 102) and $\tau_{mnkl} = 1$. Condition (8) also holds (cf. [15], p. 262, or [10], p. 315). It remains only to apply Theorems 5 and 3 of [4].

BIBLIOGRAPHY

1. N. K. Bari, *Trigonometric series*, Fizmatgiz, Moscow, 1961; English transl., Vols. I, II, Pergamon Press, Oxford, and Macmillan, New York, 1964.

2. S. Baron, *Summation multipliers for double series which are summable or bounded by Cesàro methods of real order*, Tartu Riikl Ül. Toimetised No. 102 (1961), 91–117. (Russian)

3. _____, *Summation multipliers for double series which are summable or bounded by the method of weighted Riesz means*, Tartu Riikl Ül. Toimetised No. 129 (1962), 225–240. (Russian)

4. _____, *On the extension of the Moore-Kangro method to double series*, Izv. Vyssh. Uchebn. Zaved. Mat. **1971**, no. 7 (110), 20–31. (Russian)

5. _____, *Introduction to the theory of summability of series*, 2nd ed., "Valgus", Tallinn, 1977. (Russian)

6. S. Baron and M. Tyakht [Täht], *Summation multipliers of double series for the methods $A^{\alpha,\beta}$*, Tartu Riikl Ül. Toimetised No. 305 (1972), 185–198. (Russian)

7. F. Vikhmann [Vichmann], *Theorems of Bohr-Hardy type for double series*, Tartu Riikl Ül. Toimetised No. 129 (1962), 194–198. (Russian)

8. V. K. Dzyadyk, *Introduction to the theory of uniform approximation of functions by polynomials*, "Nauka", Moscow, 1977. (Russian)

9. L. V. Zhizhiashvili, *Conjugate functions and trigonometric series*, Izdat. Tbilis. Univ., Tbilisi, 1969. (Russian)

10. A. Zygmund, *Trigonometric series*, 2nd rev. ed., Vol. I, Cambridge Univ. Press, 1959.

11. ____, *Trigonometric series*, 2nd rev. ed., Vol. II, Cambridge Univ. Press, 1959.

12. G. Kangro, *On summation multipliers for double series*, Tartu Riikl. Ül. Toimetised No. 46 (1957), 3–42. (Russian)

13. L. V. Kantorovich and G. P. Akilov, *Functional analysis*, 2nd rev. ed., "Nauka", Moscow, 1977; English transl., Pergamon Press, 1982.

14. I. G. Kull', *Multiplication of summable double series*, Tartu Riikl. Ül. Toimetised No. 62 (1958), 3–59. (Russian)

15. S. M. Nikol'skiĭ, *On linear methods of summing Fourier series*, Izv. Akad. Nauk SSSR Ser. Mat. **12** (1948), 259–278. (Russian)

16. Frédéric [Frigyes] Riesz and Béla Sz.-Nagy, *Leçons d'analyse fonctionnelle*, 2nd ed., Akad. Kiadó, Budapest, 1953; English transl., Ungar, New York, 1955.

17. M. G. Skvortsova, *On the theory of multipliers which transform double Fourier series. I*, Kabardino-Balkarsk. Gos. Univ. Uchen. Zap. No. 2 (1957), 219–231. (Russian)

18. ____, *On the theory of multipliers which transform double Fourier series. II*, Kabardino-Balkarsk. Gos. Univ. Uchen. Zap. No. 16 (1962), 40–42. (Russian)

19. ____, *On the theory of multipliers which transform double Fourier series. III*, Kabardino-Balkarsk Gos. Univ. Uchen. Zap. No. 17 (1963), 19–22. (Russian)

20. V. I. Smirnov, *A course in higher mathematics*. Vol. V, rev. ed., Fizmatgiz, Moscow, 1959; English transl., Pergamon Press, Oxford, and Addison-Wesley, Reading, Mass., 1964.

21. Elias M. Stein and Guido Weiss, *Introduction to Fourier analysis on Euclidean spaces*, Princeton Univ. Press, Princeton, N. J., 1971.

22. A. F. Timan, *Theory of approximation of functions of a real variable*, Fizmatgiz, Moscow, 1960; English transl., Pergamon Press, Oxford, and Macmillan, New York, 1963.

23. G. P. Tolstov, *Fourier series*, 2nd ed., Fizmatgiz, Moscow, 1960; German transl. of 1st ed., VEB Deutscher Verlag Wiss., Berlin, 1955; English transl. of 1st ed., Prentice-Hall, Englewood Cliffs, N. J., 1962.

24. M. Tynnov [Tönnov], *Summation multipliers, Fourier coefficients, and multipliers*, Tartu Riikl. Ül. Toimetised No. 192 (1966), 82–97. (Russian)

25. G. Goes, *Charackterisierung von Fourierkoeffizienten mit einem Summierbarkeitsfaktorentheorem und Multiplikatoren*, Studia Math. **19** (1960), 133–148.

Translated by R. L. COOKE

Amer. Math. Soc. Transl.
(2) Vol. **133**, 1986

On the Conjugation Operator in Spaces of Hölder Type
UDC 517.983:517.968

M. Z. BERKOLAĬKO

1. In this note the conjugation operator

$$(Gf)(t) = \tilde{f}(t) = -\frac{1}{2\pi}\text{P.V.}\int_{-\pi}^{\pi} f(s)\cot\frac{s-t}{2}\,ds$$

is regarded as a linear operator in a certain family of spaces of continuous 2π-periodic functions.

Denote by Φ the class of continuous increasing functions $\varphi(s)$ ($\varphi(0) = 0$) on $[0, \pi]$ such that $\varphi(s) \cdot s^{-1}$ is almost decreasing.

It follows from the Bari-Stechkin theorem [1], in particular, that if the space

$$H_\varphi = \left\{ f \in C_{[-\pi,\pi]} : \|f\|_\varphi = \|f\|_C + \sup_{0 < s \leq \pi} \frac{\omega(f, s)}{\varphi(s)} < \infty \right\}$$

is invariant with respect to G, then there exist constants $0 < \alpha < \beta < 1$ such that ($H_\alpha = H_\varphi$ for $\varphi(s) = s^\alpha$)

$$H_\beta \subsetneq H_\varphi \subsetneq H_\alpha. \tag{1}$$

Spaces different from H_φ and defined by integral metrics were considered in [2]–[4], but the sufficient conditions for them to be invariant under G are such that imbeddings of the type (1) are valid.

At the same time, it is known that there are the *Zygmund* spaces [5], which are defined in terms of the second-order modulus of continuity, invariant under G, and imbedded in all the spaces H_β ($0 < \beta < 1$).

Thus, there arise two natural questions:

1) Is there a Banach space E which does not coincide with any of the spaces H_φ ($\varphi \in \Phi$) but inherits the main property of these spaces ($\omega(f, s) \leq \omega(g, s)$ & $g \in H_\varphi \Rightarrow f \in H_\varphi$), is invariant under G, and is such that ($H_1 = H_\alpha$ for $\alpha = 1$)([1])

$$H_1 \subsetneq E \subsetneq H_\beta \qquad \forall \beta \in (0,1)? \tag{2}$$

1980 *Mathematics Subject Classification* (1985 *Revision*). Primary 26A16, 42A50, 46E35.
Translation of Soobshch. Akad. Nauk Gruzin. SSR **99** (1980), 281–284; MR **82g**:45009.
([1])There exist such spaces which are not invariant under G. Examples are given below.

©1986 American Mathematical Society
0065-9290/86 $1.00 + $.25 per page

2) Does there exist a space of this kind that is not imbedded in any of the spaces H_α $(0 < \alpha < 1)$?

Our note answers these questions.

2. DEFINITION 1. A space of continuous 2π-periodic functions is called an *ω-ideal space* if the norm in it is given as $\|f\|_E = \|f\|_C + p(f)$, where $p(f)$ is a seminorm which is monotone with respect to the modulus of continuity: $\omega(f,s) \le \omega(g,s) \Rightarrow p(f) \le p(g)$.

DEFINITION 2. An ω-ideal space E is called a *space of Hölder type* if $(n \to \infty)$
(i) $p(f_n) \to 0$, $f_n \in E \Rightarrow \omega(f_n, s) \to 0$ in measure, and
(ii) if $f_n \to 0$ in measure and $\|f_n\|_E \le M < \infty$, then $\|f_n\|_C \to 0$.

The spaces H_φ are of Hölder type. Suppose now that F is an ideal space on $[0, \pi]$ [6] and that $1/\varphi \notin F$. Let $H_{\varphi,F}$ denote the ω-ideal spaces with $p(f) = \|\omega(f,\cdot)/\varphi(\cdot)\|_F$.

If $F = L_p$ $(1 \le p < \infty)$, then $H_{\varphi,F} = H_{\varphi,p}$ [4]. The spaces $H_{\varphi,F}$ are of Hölder type: this follows from [7] and the next assertion.

THEOREM 1. *For an ω-ideal space E to be a space of Hölder type it is necessary, and also sufficient if* (i) *holds, that the imbedding $E \hookrightarrow C$ be compact.*

Suppose that for small s the function $\psi_1(s)$ is defined as $s(1 - \ln s)$, and the function $\psi_2(s)$ as $\ln^{-1}(1/s)$. Further, let Λ_{ψ_i} be the Lorentz spaces on $[0, \pi]$ [6] generated by the fundamental functions $\psi_i(s)$, and let $\phi_0(s) = s$. It is not hard to verify that $H_\alpha \hookrightarrow H_{\varphi_0, \Lambda_{\psi_2}}$ and $H_{\varphi_0, \Lambda_{\psi_2}} \hookrightarrow H_\alpha$ $(\forall \alpha \in (0, 1))$. Moreover, it follows from [7] that $H_{\varphi_0, \Lambda_{\psi_i}} \ne H_\varphi$ $(\forall \varphi \in \Phi)$.

3. We present the main results.

THEOREM 2. *There does not exist an ω-ideal space E that is invariant under G and such that the imbeddings* (2) *are valid.*

Let B_E be the unit ball of the space E, and let $\Omega(s) = \sup_{f \in B_E} \omega(f, s)$. Under certain additional assumptions the assertion of the preceding theorem can be refined somewhat.

THEOREM 3. *Suppose that $(\overline{B_E})_C = B_E$ and $\int_0 \Omega(s) \cdot s^{-1} ds < \infty$. If $G: E \to E$, then there is a β with $0 < \beta < 1$ such that $H_\beta \subset E$.*

THEOREM 4. *Suppose that the function $\varphi(t) \cdot t^{-1/p}$ $(\varphi \in \Phi; 1 < p < \infty)$ is almost increasing and $\varphi(t) \cdot t^{-1} \in L_q$ $(1/p + 1/q = 1)$. The following assertions are equivalent:*

a) There exists an $\varepsilon > 0$ such that the function $\varphi(t) \cdot t^{-1/p-\varepsilon}$ is almost increasing.[2]

b) $G: H_{\varphi,p} \to H_{\varphi,p}$.
See [3] and [4] about the implication a) $\Rightarrow$ b). It also follows from [4] that $H_{\varphi,p} \hookrightarrow H_\varepsilon$ in the class of spaces $H_{\varphi,p}$ $(\varphi \in \Phi; 1 < p < \infty)$; this gives a negative answer to question 2).

---

[2]It is clear that such a function automatically satisfies the conditions of the theorem.

We say that a function $f \in C$ has an *essentially nonpower modulus of continuity* (ENMC) if

$$\lim_{s \to 0} \int_0^s \frac{\omega(f, \varepsilon)}{\varepsilon} \, d\varepsilon / \omega(f, s) = \infty.$$

For example, functions whose moduli of continuity coincide with $\ln^{-\beta} 1/s$ ($\beta > 0$) for small s are such functions. More generally, a function $f \in C$ has an ENMC if the functions $\omega(f, s) \cdot s^{-\alpha}$ decrease for $s \in (0, s_\alpha)$ ($\forall \alpha \in (0, 1)$).

THEOREM 5. *If a function f has an ENMC, then there does not exist a space of Hölder type simultaneously containing f and invariant under G.*

REMARK. It is easy to give an example of a space E of Hölder type and a function $f_0 \in E$ having ENMC but such that $G f_0 \in E$. Nevertheless, $G(E) \not\subseteq E$ by Theorem 5. In particular, we can set $E = H_{\varphi_0, 2}$, where $\varphi_0(s)$ coincides with $\ln^{-2}(1/s)$ for small s, and we can take $f_0(t)$ to be an even increasing concave function on $[0, \pi]$ that coincides with φ_0 for small t.

BIBLIOGRAPHY

1. N. K. Bari and S. B. Stechkin, *Best approximations and differential properties of two conjugate functions*, Trudy Moskov. Mat. Obshch. **5** (1956), 483–522.
2. A. S. Dzhafarov, *Trigonometric conjugate functions and one of their connections with singular integral equations*, Izv. Akad. Nauk Azerbaĭdzhan. SSR Ser. Fiz.-Tekhn. Mat. Nauk **1974**, no. 2, 21–26. (Russian)
3. Kh. Sh. Mukhtarov, *Some multiplicative inequalities and their application to linear singular integral operators*, Dokl. Akad. Nauk SSSR **182** (1968), 764–767; English transl. in Soviet Math. Dokl. **9** (1968).
4. M. Z. Berkolaĭko, *Estimates of the moduli of continuity of functions in the spaces $B_{p,\theta}^{\alpha,\varphi}$ and their applications*, Dokl. Akad. Nauk SSSR **233** (1977), no. 5, 761–764; English transl. in Soviet Math. Dokl. **18** (1977).
5. A. Zygmund, *Trigonometric series*, 2nd rev. ed., Vol. I, Cambridge Univ. Press, 1959.
6. S. G. Kreĭn, Yu. I. Petunin and E. M. Semenov, *Interpolation of linear operators*, "Nauka", Moscow, 1978; English transl., Amer. Math. Soc., Providence, R.I., 1982.
7. M. Z. Berkolaĭko and Ya. B. Rutitskiĭ, *On a class of function spaces*, Dokl. Akad. Nauk SSSR **244** (1979), 16–19; English transl. in Soviet Math. Dokl. **20** (1979).

Translated by H. H. McFADEN

Amer. Math. Soc. Transl.
(2) Vol. **133**, 1986

Some Optimization Problems
Connected with the Construction of Stable
Dynamical Systems by Lyapunov's Second Method

E. G. ANTSIFEROV

This paper treats the problem of choosing parameters in the coefficient matrix of a linear differential system so as to ensure asymptotic stability of the system. Similar problems have been treated in [1], [2], and [13]–[16]. The basic method of solution consists in a preliminary reduction of the given problem to a linear programming problem which is then solved by the methods of linear programming theory. For example, in [16] the reduction is effected by using the Vieta formulas relating the coefficients of a characteristic polynomial to the eigenvalues. In the present paper it is based on a theorem from Lyapunov's second method.

This theorem enables one to formulate two linear programming problems equivalent to the original one. In this paper, we describe the technical details of the solution of these problems, several algorithms which employ this technique, and the solution of the matrix Lyapunov equation in the case that the coefficient matrix is presented in the Frobenius canonical form. This then provides a sufficient condition for stability of a polynomial, and an algorithm for the original problem, employing this condition.

§1. Formulation of the problem

Consider the system of ordinary differential equations

$$\dot{x} = A(u)x, \tag{1}$$

where $A(u)$ is a real n-square matrix with entries depending on a vector of parameters $u \in U \subset E_m$.

1980 *Mathematics Subject Classification* (1985 *Revision*). Primary 34C35, 34D20, 65K05.
Translation of Methods of Optimization and Their Applications (A. P. Merenkov and V. P. Bulatov, editors), "Nauka", Novosibirsk, 1982, pp. 3–22; MR **84g**:93055.

©1986 American Mathematical Society
0065-9290/86 $1.00 + $.25 per page

Throughout this paper we assume that this dependence has the form

$$A(u) = A_0 + \sum_{i=1}^{m} u_i A_i \tag{2}$$

with given n-square matrices A_i, $i = 0, 1, \ldots, m$. We also assume that the set U of admissible parameter values is an m-dimensional parallelepiped,

$$U = \{u \colon u_i^- \le u_i \le u_i^+, \ i = 1, \ldots, m\}. \tag{3}$$

Problem (A) consists of determining those parameter values $u \in U$ for which $A(u)$ becomes a stability matrix.

§2. Two basic programming problems engendered by the Lyapunov theorem

A theorem of Lyapunov's second method states that a matrix A is a stability matrix if and only if there exists a positive definite matrix X such that the matrix

$$S = XA + A^T X \tag{4}$$

is negative definite.

Let $S(X, u)$ denote the symmetric matrix defined by

$$S(X, u) = XA(u) + A^T(u)X. \tag{5}$$

The theorem just mentioned makes it possible to reformulate Problem (A) as follows:

To find a positive definite symmetric n-square matrix X and a parameter vector $u \in U$ in such a way that the matrix $S(X, u)$ defined by (5) becomes negative definite. Thus, in addition to having variable $u \in U$, one may choose arbitrarily the positive definite symmetric matrix X.

Let us introduce some notation. For any n-square symmetric matrices set

$$(C, D) = \sum_{i=1}^{n} \sum_{j=1}^{n} c_{ij} d_{ij}, \qquad \|C\|^2 = (C, C),$$

$$\lambda_-^C = \min_{\|y\|=1} (y, Cy), \qquad \lambda_+^C = \max_{\|y\|=1} (y, Cy),$$

the latter denoting the least and the largest eigenvalues of C, $y \in E_n$; and

$$F(X, Z, u) = S(X, u) + Z$$

where Z is a symmetric n-square matrix.

Let $R(\alpha, \beta)$ be the set of all symmetric matrices all of whose eigenvalues lie in the segment $[\alpha, \beta]$, i.e.,

$$R(\alpha, \beta) = \{X \colon \lambda_+^X \le \beta, \ \lambda_-^X \ge \alpha\};$$

$$\varphi(X, Z, u) = \|S(X, u) + Z\|^2 = \|F(X, Z, u)\|^2;$$

$$\psi(X, u) = \lambda_+^{S(X, u)} = \max_{\|y\|=1} (y, S(X, u)y).$$

With this notation, Problem (A) may be formulated in two ways in mathematical programming terms.

PROBLEM ψ. Minimize $\psi(X, u)$ subject to

$$X \in R(\alpha, \beta), \qquad u \in U.$$

REMARK 1. If in Problem ψ admissible values X^* and u^* are found providing a negative value of the function ψ, then Problem (A) is solved because of Lyapunov's theorem, and the matrix $A(u^*)$ is stable.

PROBLEM φ. Minimize $\varphi(X, Z, u)$ subject to

$$X \in R(\alpha, \infty), \qquad Z \in R(\beta, \infty), \quad u \in U.$$

In either problem, α and β are small positive numbers selected for each specific situation.

We shall have occasion to use the Wielandt-Hoffman inequality

$$\|A - B\|^2 \geq \sum_{i=1}^{n} (\mu_i - \nu_i)^2, \tag{6}$$

where $\mu_1 \leq \mu_2 \leq \cdots \leq \mu_n$ and $\nu_1 \leq \nu_2 \leq \cdots \leq \nu_n$ are the eigenvalues of A and B; cf. [3].

A property of Problems φ and ψ is formulated in the following lemma.

LEMMA 1. *The functions $\varphi(X, Z, u)$ and $\psi(X, u)$ are convex in each variable separately; the set $R(\alpha, \beta)$ is convex for all α and β.*

PROOF. From the definitions, $\varphi(X, Z, u)$ is a convex quadratic functional separately in each variable X, Z, u. Let us show that $R(\alpha, \beta)$ is a convex set. If $X_1 \in R(\alpha, \beta)$, $X_2 \in R(\alpha, \beta)$, and $X = (X_1 + X_2)/2$, then

$$\min_{\|y\|=1} (y, X_1 y) \geq \alpha, \qquad \min_{\|y\|=1} (y, X_2 y) \leq \alpha,$$

$$\min_{\|y\|=1} (y, Xy) \geq \frac{1}{2} \left\{ \min_{\|y\|=1} (y, X_1 y) + \min_{\|y\|=1} (y, X_2 y) \right\} \geq \alpha.$$

Analogously one shows that $\max_{\|y\|=1}(y, Xy) \leq \beta$, so that $X \in R(\alpha, \beta)$. Thus $R(\alpha, \beta)$ is convex. To prove that ψ is convex in X and u separately we express $\psi(X, u)$ in two ways,

$$\psi(X, u) = \max_{\|y\|=1} (yy^T A^T(u) + A(u)yy^T, X),$$

$$\psi(X, u) = 2 \max_{\|y\|=1} \left(Xyy^T, A_0 + \sum_{i=1}^{m} u_i A_i \right).$$

With the notation

$$C(y, u) = yy^T A^T(u) + A(u)yy^T, \tag{7}$$

$$c_i(y, X) = 2(Xyy^T, A_i), \qquad i = 0, 1, \ldots, m, \tag{8}$$

we have

$$\psi(X, u) = \max_{\|y\|=1} (C(y, u), X),\tag{9}$$

$$\psi(X, u) = \max_{\|y\|=1} \left\{ c_0(y, X) + \sum_{i=1}^{m} u_i c_i(y, X) \right\}.\tag{10}$$

Convexity of ψ is now obvious from (9) and (10), since ψ has been expressed as the supremum of continuum-many linear forms. This concludes the proof.

In the sequel the following notation will be useful:

$$S_0(X, Z) = XA_0 + A_0^T X + Z,$$
$$S_i(X) = XA_i + A_i^T X, \qquad i = 1, \ldots, m.\tag{11}$$

§3. The construction of elementary operations for solving problems φ and ψ

In the solution of nonlinear programming problems various algorithms are used. Often these algorithms require several elementary operations such as the determination of gradients or support functionals, projection onto an admissible set, determination of an optimal or admissible step in given direction, etc.

Using the notation of the preceding section, we shall now present such operations for Problems φ and ψ.

$1°$. *The partial gradient of φ with respect to X:*

$$\frac{1}{2}\frac{\partial \varphi}{\partial x_{ts}} = \sum_{i=1}^{n}\sum_{j=1}^{n} f_{ij}\frac{\partial_{ij}}{\partial x_{ts}}, \qquad f_{ij} = \sum_{k=1}^{n}(x_{ik}a_{kj} + x_{jk}a_{ki}) + z_{ij}.$$

$$\frac{\partial f_{ij}}{\partial x_{ts}} = \begin{cases} 0, & \text{if } i \neq t, \ j \neq s, \\ a_{ti}, & \text{if } i \neq t, \ j = s, \\ a_{sj}, & \text{if } i = t, \ j \neq s, \\ a_{ss} + a_{tt}, & \text{if } i = t, \ j = s. \end{cases}$$

$$\frac{1}{2}\frac{\partial \phi}{\partial x_{ts}} = \sum_{i \neq t}\sum_{j=1}^{n} f_{ij}\frac{\partial f_{ij}}{\partial x_{ts}} + \sum_{j=1}^{n} f_{tj}\frac{\partial f_{tj}}{\partial x_{ts}}$$

$$= \sum_{i \neq t}\sum_{j \neq s} f_{ij}\frac{\partial f_{ij}}{\partial x_{ts}} + \sum_{i \neq t} f_{is}\frac{\partial f_{is}}{\partial x_{ts}} + \sum_{j \neq s} f_{tj}\frac{\partial f_{tj}}{\partial x_{ts}} + f_{ts}\frac{\partial f_{ts}}{\partial x_{ts}}$$

$$= 0 + \sum_{i \neq t} f_{is}a_{ti} + \sum_{j \neq s} f_{tj}a_{sj} + f_{ts}(a_{ss} + a_{tt})$$

$$= \sum_{i=1}^{n}(a_{ti}f_{is} + f_{ti}a_{si}).$$

In matrix form, this becomes

$$\tfrac{1}{2}\nabla_X \varphi(X, Z, u) = AF + FA^T.\tag{12}$$

$2°$. *The partial gradient of φ with respect to Z:*

$$\tfrac{1}{2}\nabla_Z\varphi(X, Z, u) = F(X, Z, u). \tag{13}$$

$3°$. *The partial gradient of φ with respect to u.* With the notation (11) the function may be expressed as

$$\varphi = \left\| S_0 + \sum_{i=1}^{m} u_i S_i \right\|^2 = (u, Ru) + 2(b, u) + \varphi_0. \tag{14}$$

Here R is a symmetric m-square matrix with entries $v_{ij} = (S_i, S_j)$, b is an m-vector with components $b_i = (S_0, S_i)$, and $\varphi_0 = \|S_0\|^2$. From (14) there now follows

$$\tfrac{1}{2}\nabla_u\varphi(X, Z, u) = Ru + b. \tag{15}$$

Let $\bar{y}$ be an eigenvector corresponding to the largest eigenvalue of the matrix $S(\overline{X}, \bar{u})$.

$4°$. *The support functional of φ relative to X at a point $(\overline{X}, \bar{u})$:*

$$\varphi(X, \bar{u}) - \varphi(\overline{X}, \bar{u}) = \max_{\|y\|=1} (C(y, \bar{u}), X) - (C(\bar{y}, \bar{u}), \overline{X})$$

$$\geq (C(\bar{y}, \bar{u}), X - \overline{X}).$$

Since X is arbitrary, we have

$$\tfrac{1}{2}\hat{\nabla}_X\psi(X, u)|_{\overline{X}, \bar{u}} = \tfrac{1}{2}C(\bar{y}, \bar{u}) = \tfrac{1}{2}(\bar{y}\,\bar{y}^T A^T(\bar{u}) + A(\bar{u})\bar{y}\,\bar{y}^T). \tag{16}$$

$5°$. *The support functional of ψ relative to u at a point $(\overline{X}, \bar{u})$:*

$$\psi(\overline{X}, u) - \psi(\overline{X}, \bar{u}) = \max_{\|y\|=1} \left\{ c_0(y, \overline{X}) + \sum_{i=1}^{m} u_i c_i(y, \overline{X}) \right\}$$

$$- \left\{ c_0(\bar{y}, \overline{X}) + \sum_{i=1}^{m} \bar{u}_i c_i(\bar{y}, \overline{X}) \right\}$$

$$\geq \sum_{i=1}^{m} c_i(\bar{y}, \overline{X})(u_i - \bar{u}_i).$$

Since u is arbitrary, we have

$$\left(\tfrac{1}{2}\hat{\nabla}_u\psi(X, u)|_{\overline{X}, \bar{u}} \right)_i = \tfrac{1}{2}c_i(\bar{y}, \overline{X}) = (\overline{X}\bar{y}\bar{y}^T, A_i), \qquad i = 1, \ldots, m. \tag{17}$$

$6°$. *Projection of the symmetric matrix X onto the set $R(\alpha, \beta)$.* Let $X \notin R(\alpha, \beta)$, with $\lambda_1 < \alpha$, ..., $\lambda_k < \alpha$, $\lambda_{k+1} \in [\alpha, \beta]$, ..., $\lambda_s \in [\alpha, \beta]$, $\lambda_{s+1} > \beta$, ..., $\lambda_n > \beta$ for the eigenvalues of X, and let $v_1, \ldots, v_n$ be an orthonormal system of corresponding eigenvectors; then $V = (v_1, \ldots, v_n)$ is an orthogonal matrix such that

$$V^T X V = \operatorname{diag}\{\lambda_i\} = \Lambda.$$

Set up the matrix $\overline{\Lambda} = \operatorname{diag}(\bar{\lambda}_i)$ by letting

$$\bar{\lambda}_i = \begin{cases} \alpha, & \text{if } \lambda_i < \alpha, \\ \lambda_i, & \text{if } \lambda_i \in [\alpha, \beta], \\ \beta, & \text{if } \lambda_i > \beta. \end{cases}$$

Let $\overline{X} = V\overline{\Lambda}V^T$, and let Y be any matrix in $R(\alpha, \beta)$ with eigenvalues $\mu_1, \ldots, \mu_n$. By the Wielandt-Hoffman inequality (6),

$$\|X - Y\|^2 \geq \sum_{i=1}^{n}(\lambda_i - \mu_i)^2 \geq \min_{\alpha \leq \mu_i \leq \beta} \sum_{i=1}^{n}(\lambda_i - \mu_i)^2 = \sum_{i=1}^{n}(\lambda_i - \bar{\lambda}_i)^2$$

$$= \|\Lambda - \overline{\Lambda}\|^2.$$

Since orthogonal transformations preserve Euclidean lengths,

$$\|X - Y\|^2 \geq \|V(\Lambda - \overline{\Lambda})V^T\|^2 = \|X - \overline{X}\|^2, \tag{18}$$

Since $\overline{X} \in R(\alpha, \beta)$ and Y is arbitrary in $R(\alpha, \beta)$, inequality (18) yields that $\overline{X}$ is the projection of X onto $R(\alpha, \beta)$. This projection of X onto $R(\alpha, \beta)$ will be denoted by $\pi_{\alpha,\beta}(X)$. In actual realization it is not necessary to determine all the eigenvalues and eigenvectors of X. It suffices to use the formula

$$\pi_{\alpha,\beta}(X) = X + \sum_{k \in I_\alpha}(\alpha - \lambda_k)v_k v_k^T + \sum_{k \in J_\beta}(\beta - \lambda_k)v_k v_k^T. \tag{19}$$

Here $I_\alpha = \{k\colon \lambda_k < \alpha\}$ and $J_\beta = \{k\colon \lambda_k > \beta\}$. The advantage of (19) is in that one may successively determine the $(\lambda_1, v_1), \ldots$ and then, when $k \in I_\alpha \cup J_\beta$, add the needed components to X according to (19). In the projection procedure one would prefer methods which determine the eigendata (λ_k, v_k) of given index k in the ordering $\lambda_1 \leq \lambda_2 \leq \cdots \leq \lambda_n$, for example, the method due to Givens and Householder [3].

7°. *Admissible step size in a given direction.* Let $X_0 \in R(\alpha, \beta)$ and let P be a symmetric matrix. Consider the one-parameter family of matrices

$$X(t) = X_0 + tP, \qquad t \geq 0.$$

The question is to determine admissible step sizes t^* in the direction of P, i.e., to find $\max t$ satisfying the conditions $t \geq 0$ and $X(t) \in R(\alpha, \beta)$.

Let v_-^P and v_+^P be eigenvectors of P corresponding to λ_-^P and λ_+^P. Then, for each admissible t,

$$(v_-^P, X(t)v_-^P) = (v_-^P, X_0 v_-^P) + \lambda_-^P t \geq \alpha. \tag{20}$$

If $\lambda_-^P \geq 0$ then (20) is satisfied for all $t \geq 0$; in the remaining case,

$$t^* \leq \bar{t}_1 = [(v_-^P, X_0 v_-^P) - \alpha]/(-\lambda_-^P). \tag{21}$$

On the other hand, for any admissible t,

$$(v_+^P, X_0 v_+^P) + \lambda_+^P t \leq \beta. \tag{22}$$

If $\lambda_+^P \leq 0$, then (22) is satisfied for all $t \geq 0$; in the remaining case,

$$t^* \leq \bar{t}_2 = [\beta - (v_+^P, X_0 v_+^P)]/\lambda_+^P. \tag{23}$$

We thus obtain an upper estimate for t^*:

$$t^* \leq t_0 = \min\{\bar{t}_1, \bar{t}_2\},$$

where

$$\bar{t}_1 = \begin{cases} [(v_-^P, X_0 v_-^P) - \alpha]/(-\lambda_-^P), & \text{if } \lambda_-^P < 0, \\ \infty, & \text{if } \lambda_-^P \geq 0; \end{cases}$$

$$\bar{t}_2 = \begin{cases} [\beta - (v_+^P, X_0 v_+^P)]/\lambda_+^P, & \text{if } \lambda_+^P > 0, \\ \infty, & \text{if } \lambda_+^P \leq 0. \end{cases}$$

It may be noted that $t_0 = \infty$ only when $P = 0$.

§4. Algorithmic solution of Problems φ and ψ

In this section we shall assume that the matrix Z is fixed, e.g., $Z = I$, the identity matrix. From Lemma 1 it follows that Problems φ and ψ with fixed u (or fixed X) are convex programming problems. The theory and solution methods for such problems have been well developed. It then appears natural to apply, for Problems φ and ψ, group relaxation methods, with respect to two groups of variables: $u \in U$ and $X \in R(\alpha, \beta)$.

For notational convenience let us introduce the following identifiers: $A\Phi XK$, $A\Phi UK$, $A\Psi XK$, and $A\Psi UK$, where K denotes a positive integer. These will serve to denote various solution algorithms for Problems φ and ψ. The initial letter A corresponds to the word "algorithm". If the second letter is Φ, the algorithm solves Problem φ; otherwise it solves Problem ψ. If the third letter is X, the minimization is over $X \in R(\alpha, \beta)$; otherwise it is over $u \in U$. Finally, K is the algorithm number.

The algorithms $A\Phi UK$ minimize the convex quadratic form (14) with parallelepiped constraint sets. Among several known algorithms of this class we mention that due to Polyak [4], a modification of the congruent gradient method. In actual calculations we have used another one, due to Dikin [5], the method of interior points.

AΦU1. We begin at an interior point u^0 of the parallellepiped, $u^- < u^0 < u^+$. The sequence u^k is then constructed as follows (notation from (11), (14), and (15)):

1. One computes $c^k = \frac{1}{2}\nabla_u \varphi_k = Ru_k + b$.
2. An auxiliary elementary problem is solved:

$$\min \left\{ (c^k, u) \Big/ \sum_{i=1}^{m} (\sigma_i^k)^2 (u_j - u_i^k)^2 = 1 \right\},$$

where

$$1/(\sigma_i^k)^2 = [(u_i^k - u_i^-)^2 \times (u_i^k - u_i^+)^2]/[(u_i^k - u_i^-)^2 + (u_i^k - u_i^+)^2].$$

The solution $\tilde{u}_k$ of the auxiliary problem is easily found via Lagrange multipliers; one then determines u^{k+1} by minimizing a quadratic form on the interval $[u^k, \tilde{u}^k]$.

In [5] it is shown that the sequence of inner points u^k converges to a solution of the problem. In practical applications, $A\Phi U1$ exhibits good convergence properties.

AΦX1. First note that every symmetric matrix $X \in R(\alpha, \infty), \alpha > 0$, may be expressed as

$$X = \alpha I + Y^T Y, \tag{24}$$

with Y upper triangular. This leads to the problem of finding an absolute minimum of

$$\varphi_1(Y) = \|Y^T Y \overline{A} + \overline{A}^T Y^T Y + I_\alpha\|^2, \tag{25}$$

with $\overline{A} = A(\bar{u})$ and $I_\alpha = I + \alpha(\overline{A} + \overline{A}^T)$.

The constraint $X \in R(\alpha, \infty)$ is dropped, and the functional is no longer convex. The conjugate gradient method is used to minimize $\varphi_1(Y)$, with global minimization in the direction chosen. Let us describe one step of the algorithm. For given Y^k one determines:

1. $X^k = \alpha I + (Y^k)^T Y^k$.
2. $F^k = F(X^k, I_\alpha, \bar{u})$, $\varphi_k = \|F^k\|^2$.
3. $\nabla \varphi_1^k$ is the upper triangle of the matrix $Y^k[F^k \overline{A}^T + \overline{A} F^k]$.
4. $P^k = -\nabla \varphi_1^k + \beta_k P^{k-1}$, where $\beta_k = 0$ when $k = 0$, and for $k \neq 0$, $\beta_k = \|\nabla \varphi_1^k\|^2 / \|\nabla \varphi_1^{k-1}\|^2$. In the direction of decrease of P^k we take the h-step

$$X(h) = X^k + [(Y^k)^T P^k + (P^k)^T Y^k]h + (P^k)^T P^k h^2$$
$$= X^k + hX_1^k + h^2 X_2^k;$$

$$F(h) = F^k + [X_1^k \overline{A} + \overline{A}^T X_1^k]h^2 + [X_2^k \overline{A} + \overline{A}^T X_2^k]h^2$$
$$= F^k + hF_1^k + h^2 F_2^k;$$

$$\varphi_1(Y^k + hP^k) = \varphi_k + hc_{1k} + h^2 c_{2k} + h^3 c_{3k} + h^4 c_{4k} = \chi(h);$$

$$c_{1k} = 2(F^k, F_1^k); \qquad c_{2k} = \|F_1^k\|^2 + 2(F^k, F_2^k);$$

$$c_{3k} = 2(F_1^k, F_2^k); \qquad c_{4k} = \|F_2^k\|^2.$$

5. The minimum h_k^* of $\chi^{(h)}$ is found as the real root of the equation

$$\chi'(h) = c_{1k} + 2hc_{2k} + 3h^2 c_{3k} + 4h^3 c_{4k} = 0.$$

6. Set $Y^{k+1} = Y^k + h_k^* P^k$ and repeat iteratively. The algorithm $A\Phi X1$ has performed well in practice, particularly in initial iterations.

$A\Phi X2$. We choose a sequence of numbers ρ_k, $k = 0, 1, \ldots$, satisfying

$$\rho_k \geq 0, \qquad \rho_k \to 0, \qquad \sum_{k=0}^{\infty} \rho_k = \infty.$$

Beginning with $X^0 \in R(\alpha, \infty)$ one constructs the sequence X^k by setting $X^{k+1} = \pi_{\alpha, \infty}(X^k - \rho_k \nabla_X \varphi(X^k, I, \bar{u}))$. Thus $A\Phi X2$ is an interpretation of an algorithm presented in [6]; it is mentioned here since we have an efficient procedure to implement projection.

In actual computations the following was also used:

$A\Phi X3$. This is a variant of the projected gradient method; it also requires a projection procedure. Let us describe its main features.

Beginning with a point $X^k \in R(\alpha, \infty)$, one determines $\overline{X}^k$ as the optimal point in the direction of greatest decrease. If then $\overline{X}^k \in R(\alpha, \infty)$ we set $X^{k+1} = \overline{X}^k$; if not, let X^{k+1} be the optimal point on the segment $[X^k, \pi_{\alpha, \infty}(\overline{X}^k)]$.

A formal description of one iteration is as follows. For given $X^k \in R(\alpha, \infty)$ one finds

1. $F^k = F(X^k, I, \bar{u})\varphi_k = \|F^k\|^2$.
2. $\nabla\varphi^k = F^k\overline{A}^T + \overline{A}F^k = -P^k$.
3. $S^k = P^k\overline{A} + \overline{A}^T P^k$.
4. $c_{1k} = (S^k, F^k)$, $c_{2k} = \|S^k\|^2$.

In the direction of greatest decrease we have

$$X(h) = X^k + hP^k, \quad F(h) = F^k + hS^k; \quad \varphi(X(h), I, \bar{u}) = \varphi_k + 2hc_{1k} + h^2 c_{2k}.$$

5. Set $h_k^* = -c_{1k}/c_{2k}$ and $X_0^k = X^k + h_k^* P^k$; if $X_0^k \in R(\alpha, \infty)$ then set $X^{k+1} = X_0^k$, and repeat the cycle.
6. $\overline{X}^k = \pi_{\alpha,\infty}(X_0^k)$.
7. $\overline{P}^k = \overline{X}^k - X^k$.
8. One determines h_k^* by minimizing $\varphi(X^k + h\overline{P}^k, I, u)$ over $h \in [0, 1]$.
9. Set $X^{k+1} = X^k + h_k^*\overline{P}^k$ and repeat the cycle.

АФХ4. *Minimization over columns of the matrix X*. Fix all entries in X except those in the first column, $x_1 = x_{11}, \ldots, x_n = x_{n1}$. Denote

$$x = (x_1, \ldots, x_n)^T, \qquad y = (x_2, \ldots, x_n)^T = (y_1, \ldots, y_{n-1})^T,$$

$$B = \begin{bmatrix} \bar{x}_{22} & \cdots & \bar{x}_{2n} \\ \cdots & \cdots & \cdots \\ \bar{x}_{n2} & \cdots & \bar{x}_{nn} \end{bmatrix}. \tag{26}$$

Barred entries indicate the entries of X that are unchanged in the iteration under consideration. With this notation, X may be written as

$$X(x) = \begin{bmatrix} x_1 & y \\ y^T & B \end{bmatrix}. \tag{27}$$

We assume that initially the matrix X was positive definite; then B is also positive definite.

According to Sylvester's criterion, $X(x)$ is positive definite if and only if

$$\det X(x) > 0 \tag{28}$$

On expanding the determinant in terms of the first row, we obtain the quadratic inequality

$$x_1 > (y, B^{-1}y). \tag{29}$$

This determines the region of admissible values for x.

It is obvious that the function φ in dependence on $x_{11}, \ldots, x_{n1}$ (i.e., as a function of x when the remaining variables are held fixed) is a convex quadratic function,

$$\varphi(X(x), I, \bar{u}) = (x, Cx) + 2(b, x) + f_0 = f(x). \tag{30}$$

For applying $A\Phi X4$ it will be useful to develop formulas for C, b, and f_0. Denote by E_j the matrix with entry 1 in the jth diagonal position and zero entries elsewhere. Let $\overline{F}_j$ be the jth row of the matrix $\overline{F} = F(X(0), I, \bar{u})$. Then

$$\begin{cases} D_1 = \overline{A}_+^T \bar{a}_{11} I + \sum_{i=2}^n \bar{a}_{i1}, \\ D_j = \bar{a}_{1j} I + \sum_{i=2}^n \bar{a}_{ij} E_i + \overline{A}^T E_j, \quad j = 2, \ldots, n \end{cases} \tag{31}$$

Then C, b, and f_0 may be expressed in terms of the D_j and $\overline{F}_j$:

$$C = \sum_{j=1}^n D_j^T D_j, \qquad b = \sum_{j=1}^n D_j^T \overline{F}_j, \qquad f_0 = \sum_{j=1}^n \|\overline{F}_j\|^2. \tag{32}$$

Let us now describe the algorithm.

The algorithm changes a given matrix into a new one by replacing one row and column (in the present case, the first) by a vector x^* determined as the solution of the following elementary convex programming problem: To minimize

$$f(x) = (x, Cx) + 2(b, x) + f_0 \tag{33}$$

under the constraint

$$g_\varepsilon(x) = x_1 - (y, B^{-1}y) - \varepsilon \geq 0 \tag{34}$$

Here the small positive number ε is a parameter of the algorithm. For simplicity let us assume that $\det C > 0$. The following operations are applied to solve problem (33), (34).

1°. One finds the point x^0 as the unconstrained minimum of $f(x)$, $x^0 = -C^{-1}b$.

2°. If it turns out that $g_\varepsilon(x^0) \geq 0$, then x^0 is a solution of (33), (34). In the remaining case there exist a uniquely determined λ^* (a Lagrange multiplier) and a vector x^* such that

$$\nabla f(x^*) = \lambda^* \nabla g_\varepsilon(x^*), \tag{35}$$

$$g_\varepsilon(x^*) = 0. \tag{36}$$

Now write C in the form

$$C = \begin{bmatrix} \gamma & c \\ c^T & \overline{C} \end{bmatrix}, \qquad \gamma = c_{11}, \quad c = (c_{21}, \ldots, c_{n1})^T.$$

From (35), (30), and (34) we obtain

$$\begin{cases} \gamma x_1 + c^T y + b_1 = \lambda/2 \\ x_1 c + \overline{C} y + \bar{b} = -\lambda B^{-1} y, \quad \bar{b} = (b_2, \ldots, b_n)^T. \end{cases} \tag{37}$$

Since $\overline{C}$ and B are positive definite and $\lambda > 0$, we have that the matrix $D(\lambda) = \overline{C} + \lambda B^{-1}$ is invertible, and

$$x_1 = [\lambda/2 - b_1 + c^T D^{-1}(\lambda)\bar{b}] \times [\gamma - c^T D^{-1}(\lambda)c]^{-1}, \tag{38}$$

$$y = -D^{-1}(\lambda)(x_1 c + \bar{b}). \tag{39}$$

$3°$. If an upper estimate $\bar{\lambda}$ for λ^* is known, the bisection method may be applied to the interval $[0, \bar{\lambda}]$ using (38) and (39) in solving the equation $g_\varepsilon(x_1(\lambda), y(\lambda)) = 0$.

Having solved problem (33), (34), we replace the first row and column (or the row and column being treated) by x^*, and proceed to treat the remaining ones. The resulting algorithm is one of group relaxation via the rows of matrix X.

The following two algorithms are based on the cutting method [7].

AΨX1. We enclose the set $R(\alpha, \beta)$ within a parallelepiped $\Pi = \{X \colon X^- \leq X \leq X^+\}$.

$1°$. For the initial point $X^0 \in \Pi$ we construct the supporting functional $\hat{\nabla}\psi_0 = \hat{\nabla}_X \psi(x^0, \bar{u})$ in accordance with (16).

$2°$. For $k = 0$ we solve the linear programming problem $(LP)_k$: minimize v subject to

$$\{X, v\} \in R_k = \{\{X, v\} \colon (\hat{\nabla}\psi^k, X) \leq v, X \in \Pi\}.$$

For $k = 0, 1, \ldots$ let $\{\overline{X}^k, v_k\}$ be the solution of problem $(LP)_k$; let y_i^k (with $i \in I_k^+$) be the set of those eigenvectors of $\overline{X}^k$ for which $(y_i^k, \overline{X}^k y_i^k) > \beta$; and let y_i^k (with $i \in I_k$) be the set of eigenvectors of $\overline{X}^k$ for which $(y_i, \overline{X}^k y_i^k) < \alpha$. Finally, set $\hat{X}^k = \pi_{\alpha, \beta}(\overline{X}^k)$ and $\psi_k = \psi(\hat{X}^k, \bar{u})$.

$3°$. Construct the set

$$R_{k+1} = \Big\{\{X, v\} \colon \{X, v\} \in \bigcap_{j=0}^k R_j, \ (y_i^k (y_i^k)^T, X) \leq \beta, \ i \in I_k^+,$$
$$(y_i^k (y_i^k)^T, X) \geq \alpha, \ i \in I_k^-, \ (\hat{\nabla}\psi(\hat{X}^k, \bar{u}), X - \hat{X}^k) + \psi_k \leq v \Big\}.$$

$4°$. We solve problem $(LP)_{k+1}$, to minimize v subject to $\{X, v\} \in R_{k+1}$. Obviously $R_0 \supset R_1 \supset \cdots \supset R_k \supset \cdots \supset R(\alpha, \beta) \times \{v \colon -\infty \leq v \leq \infty\}$. Let ψ_* be the optimal value in Problem ψ (with $u = \bar{u}$), let ψ^j be the value of ψ at the optimal point of problem $(LP)_j$, and let $\bar{\psi}_k = \min_{0 \leq j \leq k} \psi^j$. Then, for all k, we have $v_0 \leq \cdots \leq v_{k-1} \leq v_k \leq \psi_* \leq \bar{\psi}_k \leq \bar{\psi}_{k-1} \leq \cdots \leq \bar{\psi}_1$. In [8] it is shown that

$$\lim_{k \to \infty} v_k = \lim_{k \to \infty} \bar{\psi}_k = \psi_*.$$

$5°$. The procedure is terminated as soon as $\bar{\psi}_k - v_k \leq \varepsilon$, where ε denotes the desired accuracy.

Thus $A\Psi X1$ involves the solution of an infinite sequence of problems LP with strictly increasing dimensionality. This may inhibit the realization on computers with small memory. On the other hand, algorithms of this type have one undeniable advantage: they provide two-sided estimates of the optimal value of functionals with arbitrarily small error. We note that $A\Psi X1$ also uses a projection procedure to determine points of successive linearization of $\hat{X}^k$.

AΨU1. The basic idea of this algorithm is the same as for $A\Psi X1$. It differs in the method of finding support functionals (cf. (17)), and also in that the constraint set U is a parallelepiped, so that it is unnecessary to approximate it by a polyhedron.

$1°$. In accordance with (17), at the initial point $u^0 \in U$ we construct the support functional $\hat{\nabla}\psi^0 = \hat{\nabla}_u \psi(\overline{X}, u^0)$, and also the set

$$R_0 = \left\{ \{u, v\} \colon c_0(\bar{y}^0, \overline{X}) + \sum_{i=1}^m c_i(\bar{y}^0, \overline{X}) u_i \le v, \ u \in U \right\}.$$

Then we solve the linear programming problem $(LP)_0$,

$$\text{minimize } v \text{ subject to } \{u, v\} \in R_0.$$

Let $\{\bar{u}^1, v_1\}$ be the solution of this, and set $\psi_1 = \psi(\overline{X}, \bar{u}^1)$.

$2°$. For $k = 0, 1, \ldots$ we define the sets

$$R_{k+1} = \left\{ \{u, v\} \colon \{u, v\} \in \bigcap_{j=0}^k R_j, \ c_0(\bar{y}^k, \overline{X}) + \sum_{i=1}^m c_i(\bar{y}_k, \overline{X}) u_i \le v \right\}.$$

Here $\bar{y}^k$ is an eigenvector corresponding to the largest eigenvalue of the matrix $S(\overline{X}, \bar{u}^{k+1})$.

$3°$. We solve problem $(LP)_{k+1}$:

$$\text{minimize } v \text{ subject to } \{u, v\} \in R_{k+1};$$

let $\{\bar{u}^{k+2}, v_{k+2}\}$ be its solution.

$4°$. As in $A\Psi X1$, we obtain two sequences, $\bar{\psi}_k \searrow \psi_*$ and $v_k \nearrow \psi_*$, where ψ_* is the u-optimal value in Problem ψ.

REMARK 2. Suppose that, with the use of the algorithms described above or by other algorithms for minimizing φ (over X and u), one has found $\overline{X} \in R(\alpha, \infty)$ and $\bar{u} \in U$ such that $\varphi(\overline{X}, I, \bar{u}) \le 1$. Let $\bar{\sigma}_1, \ldots, \bar{\sigma}_n$ be the eigenvalues of $S(\overline{X}, \bar{u})$. According to the Wielandt-Hoffman inequality, we have $\sum(\bar{\sigma}_i + 1)^2 \le \varphi < 1$ and $\bar{\sigma}_i < 0$, $i = 1, \ldots, n$. This last shows that $S(\overline{X}, \bar{u})$ is negative definite and $A(\bar{u})$ is a stability matrix. If now we use a relaxation algorithm like $A\Phi UK$ with initial value $\bar{u}$, there will result a sequence $u^0, \bar{u}, u^1, u^2, \ldots$ such that

$$1 > \varphi(\overline{X}, I, u^0) \ge \varphi(\overline{X}, I, u^1) \ge \cdots \ge \varphi(\overline{X}, I, u^k) \ge \cdots.$$

Since φ is convex with respect to u, every point $\tilde{u}$ in the convex hull of the points $\{u^k\}$ will also generate a stable matrix $A(\tilde{u})$. This yields convex polyhedra $M \subset U$ such that $A(u)$ is stable for every $u \in M$.

If a relaxation algorithm $A\Psi UK$ is begun at $\bar{u}$, we also obtain a sequence $\{u^k\}$ for which $\{A(u^k)\}$ is a sequence of stability matrices.

§5. An algorithm for solving the Lyapunov matrix equation, and related sufficiency conditions for stability of polynomials

The literature contains a large number of algorithms for solving the Lyapunov matrix equation, i.e.

$$XA + A^T X = Z \tag{40}$$

for unknown X; here Z is a given symmetric and A a given real n-square matrix. See [9]–[11].

Let us describe one solution procedure for (40), which will be used in the sequel. A similarity transformation M is first used to bring A into Frobenius canonical form,

$$\tilde{A} = M^{-1}AM = \begin{pmatrix} 0 & 1 & 0 & \cdots & 0 & 0 \\ 0 & 0 & 1 & \cdots & 0 & 0 \\ \hdotsfor{6} \\ 0 & 0 & 0 & \cdots & 0 & 1 \\ -p_n & -p_{n-1} & -p_{n-2} & \cdots & -p_2 & -p_1 \end{pmatrix} \tag{41}$$

This may be carried out by A. M. Danilevskiĭ's method, with $O(n^3)$ as operation count. Set

$$Q = M^T Z M, \tag{42}$$

and solve the equation

$$Y\tilde{A} + \tilde{A}^T Y = Q \tag{43}$$

Let Y denote a solution of (43). Set

$$X = (M^{-1})^T Y M^{-1}. \tag{44}$$

Then

$$\begin{aligned} XA + A^T X &= (M^{-1})^T Y M^{-1} M \tilde{A} M^{-1} + (M^{-1})^T \tilde{A} M^T (M^{-1})^T Y M^{-1} \\ &= (M^{-1})^T Y \tilde{A} M^{-1} + (M^{-1})^T \tilde{A}^T Y M^{-1}. \end{aligned} \tag{45}$$

From (43) it then follows that

$$XA + A^T X = (M^{-1})^T (Y\tilde{A} + \tilde{A}^T Y) M^{-1} = (M^{-1})^T Q M^{-1} = Z \tag{46}$$

so that X is a solution of (40). Thus the solution of the Lyapunov equation is reduced to the construction of the transformation M and the solution of (43).

Let us now show that the solution of (43) reduces to the solution of a system of n equations in n unknowns relative to the nth row of Y, and a subsequent determination of the remaining entries by recurrence formulas.

From (41) and (43) we obtain

$$\begin{aligned} &\tfrac{1}{2}Q_{11} = -p_n y_{1n}, \\ &Q_{1j} = y_{1,j-1} - y_{1n}p_{n-j+1} - y_{jn}p_n, \quad i > 1, \\ &Q_{ij} = -y_{in}p_{n-j+1} - y_{jn}p_{n-i+1} + y_{i,j-1} + y_{j,i-1}, \quad i,j > 1. \end{aligned} \tag{47}$$

Our task is to obtain, from (47), a system of n equations in the unknowns $y_{1n}, \ldots, y_{nn}$. The first equation in this system has already been presented: $\tfrac{1}{2}Q_{11} = -p_n y_{1n}$. To obtain the second, we find

$$\tfrac{1}{2}Q_{22} - Q_{13} = -y_{2n}p_{n-1} + y_{12} - y_{12} + y_{1n}p_{n-2} + y_{3n}p_n$$

(using symmetry of Y). Thus the second equation is

$$\tfrac{1}{2}Q_{22} - Q_{13} = -y_{1n}p_{n-2} - y_{2n}p_{n-1} + y_{3n}p_n,$$

and involves y_{1n}, y_{2n}, and y_{3n} only. For the third equation we find

$$\tfrac{1}{2}Q_{33} - Q_{24} + Q_{15} = -y_{3n}p_{n-2} + y_{23} + y_{2n}p_{n-3} + y_{4n}p_{n-1}$$
$$- y_{23} - y_{14} + y_{14} - y_{1n}p_{n-4} - y_{5n}p_n$$

and this then has the form

$$\tfrac{1}{2}Q_{33} - Q_{24} + Q_{15} = -y_{1n}p_{n-4} + y_{2n}p_{n-3} - y_{3n}p_{n-2} + y_{4n}p_{n-1} - y_{5n}p_n.$$

For the kth equation we determine the integer s_k from the conditions $s_k + k \leq n$ and $k - s_k \geq 1$, so that $s_k = \min\{k - 1, n - k\}$. Then define

$$l_k = \tfrac{1}{2}Q_{kk} + \sum_{s=1}^{s_k}(-1)^s Q_{k-s,k+s}.$$

Direct verification yields that the terms y_{ij} $(i \neq n, j \neq n)$ do not appear in the sum l_k, and the kth equation has the form

$$\tfrac{1}{2}Q_{kk} + \sum_{s=1}^{s_k}(-1)^s Q_{k-s,k+s} = \sum_{j=1}^{n}\nu_{kj}y_{jn}, \qquad k = 1, \ldots, n, \qquad (48)$$

where

$$\nu_{kj} = \begin{cases} 0 & \text{if } j \notin [j_{kn}^-, j_{kn}^+], \\ (-1)^{j-k+1}p_{n-2k+j-1} & \text{otherwise, } k, j = 1, \ldots, n. \end{cases} \qquad (49)$$

Here $p_0 = 1$, $j_{kn}^- = \max\{1, 2k - n - 1\}$, and $j_{kn}^+ = \min\{2k - 1, n\}$.

It may be remarked that the matrix $R = \{\nu_{ij}\}$ coincides, up to signs of entries with $i - j \equiv 0 \pmod 2$, with the matrix appearing in the Routh-Hurwitz criterion. After determining $y_{1n}, \ldots, y_{nn}$ from (48) and (49), the remaining elements are found using (47); the solution of (40) is then obtained from (44). Thus the operation count for the solution of the matrix equation by this method has the order $O(n^3)$.

We formulate the preceding results in a lemma.

LEMMA 2. *Let there be given a real polynomial $P_n(\lambda) = \lambda^n + p_1\lambda^{n-1} + \cdots + p_n$ and its companion matrix*

$$A(p) = \begin{pmatrix} 0 & 1 & 0 & \cdots & 0 & 0 \\ \cdots\cdots\cdots\cdots\cdots\cdots\cdots\cdots\cdots\cdots\cdots\cdots\cdots \\ 0 & 0 & 0 & \cdots & 0 & 1 \\ -p_n & -p_{n-1} & \cdots & \cdots & -p_2 & -p_1 \end{pmatrix}.$$

Consider the Lyapunov matrix equation $XA(p) + A^T(p)X + Q = 0$ for X; to the polynomial p make correspond the modified Hurwitz matrix $R(p)$ defined by (49), and to the matrix Q the vector $f(Q)$ whose coordinates are the left-hand side in (48), equipped with a minus sign. Then, if there exists a solution of the Lyapunov equation, the nth row of X is determined from the equation

$$R(p)y = f(Q) \qquad (*)$$

This result was obtained by the American mathematician Molinari in 1969, and independently by the present author in 1971. Lemma 2 may be used to obtain sufficient conditions for stability of polynomials, and, based on this, an algorithmic solution of the basic problem (A).

Let $p^0 = (p_1^0, \ldots, p_n^0)$ be the coefficients of a stable polynomial, with companion matrix

$$A^0 = A(p^0) = \begin{pmatrix} 0 & 1 & 0 & \cdots & 0 & 0 \\ \cdots\cdots\cdots\cdots\cdots\cdots\cdots\cdots\cdots \\ 0 & 0 & 0 & \cdots & 0 & 1 \\ -p_n^0 & -p_{n-1}^0 & \cdots & \cdots & -p_2^0 & -p_1^0 \end{pmatrix}. \tag{50}$$

Consider the Lyapunov matrix equation

$$XA^0 + (A^0)^T X + Q = 0, \tag{51}$$

where Q is a positive definite matrix. According to Lemma 2, the nth column of X in (51) may be found from the equation

$$R(p^0)y = f(Q). \tag{52}$$

Let y be a solution of (52). We introduce an auxiliary vector $v = (v_1, \ldots, v_n)$, a polynomial $p(v) = (p_1^0 - v_n, p_2^0 - v_{n-1}, \ldots, p_n^0 - v_1)$, and its companion matrix $A^v = A(p(v))$. Obviously

$$A^v = A^0 + \begin{pmatrix} 0 & 0 & \cdots & 0 \\ \cdots\cdots\cdots\cdots\cdots \\ 0 & 0 & \cdots & 0 \\ v_1 & v_2 & \cdots & v_n \end{pmatrix}. \tag{53}$$

Next, consider the matrices

$$\tilde{F}(v) = X^0 V^v + (A^v)^T X^0, \qquad F(v) = \tilde{F}(v) + Q, \tag{54}$$

where X^0 is a solution of (51) and the last column of X^0 is y. From (53) and (51) it immediately follows that $F(v) = yv^T + vy^T$.

We find the square of the Euclidean norm of $F(v)$,

$$\varphi(v) = \|F(v)\|^2 = \|yv^T + vy^T\|^2 = (v, Bv). \tag{55}$$

Here B is the symmetric matrix

$$B = 2(\|y\|^2 I + yy^T). \tag{56}$$

and has the eigenvalues

$$\lambda_1 = \lambda_2 = \cdots = \lambda_{n-1} = 2\|y\|^2, \qquad \lambda_n = 4\|y\|^2. \tag{57}$$

The eigenvector of B corresponding to λ_n is collinear with y, and the remaining eigenvectors may be taken as any orthonormal basis of the subspace perpendicular to y.

The vector v has been arbitrary; let us now require that

$$(v, Bv) < (\lambda_-^Q)^2 = \lambda_{\min}^2(Q). \tag{58}$$

From (54), (55), and (58) there follows

$$\|F(v)\|^2 = \|\tilde{F}(v) + Q\|^2 < \lambda_{\min}^2(Q). \tag{59}$$

On the other hand, from the Wielandt-Hoffman lemma,

$$\sum_{i=1}^{n} (\lambda_i(\tilde{F}) + \lambda_i(Q))^2 \le \|\tilde{F}(v) + Q\|^2. \tag{60}$$

Here the $\lambda_i(\tilde{F})$ are the eigenvalues of $\tilde{F}(v)$. The conjunction of (59) and (60), together with $\lambda_i(Q) \ge \lambda_{\min}(Q)$, yields that all eigenvalues of $\tilde{F}(v)$ are negative,

$$\lambda_i(\tilde{F}) < 0, \ i = 1, \dots, n. \tag{61}$$

Thus $\tilde{F}(v)$ is negative definite for all v satisfying (58) and for the positive definite matrix X^0 found from (51). According to Lyapunov's theorem it follows that the matrix A and the polynomial $p(v)$ are stable for all v satisfying (58).

These results may be formulated in the following theorem on the ellipsoid of stable polynomials:

THEOREM 1. *Let p^0 be the coefficients of a given stable polynomial, let Q be any symmetric positive definite matrix with least eigenvalue $\lambda_{\min}(Q)$, and let B be the symmetric positive definite matrix defined by (49), (*), (52), and (56).*

Then a sufficient condition for stability of $p(v) = (p_1^0 - v_n, p_2^0 - v_{n-1}, \dots, p_n^0 - v_1)$ is that the vector $v = (v_1, \dots, v_n)$ lies within the n-dimensional ellipsoid

$$(v, Bv) < \lambda_{\min}^2(Q). \tag{$**$}$$

Since the largest eigenvalue of B is $4\|y\|^2$, the condition in the theorem will be satisfied as soon as

$$4\|y\|^2 \|v\|^2 < \lambda_{\min}^2(Q)$$

i.e.,

$$\|v\| < \lambda_{\min}(Q)/\|y\|. \tag{62}$$

From the inequality there follows

COROLLARY. *Let p^0 be a stable polynomial, and let $Q, \lambda_{\min}(Q)$, and y be as in Theorem 1. Then every polynomial within the n-dimensional ball $\|p - p^0\| < \rho$ of radius $\rho = \lambda_{\min}(Q)/(2\|y\|)$ is also stable.*

REMARK. Without loss of generality one may take $\lambda_{\min}(Q) = 1$ in (62). Indeed, if we consider two matrices related by $\tilde{Q} = kQ$, then the solutions of (62) are also related by $\tilde{y} = ky$ and $\|\tilde{y}\| = k\|y\|$, and this has no effect on the right-hand sides of (62). In connection with this there arises the problem of minimizing the function $Z(Q) = \|y(Q)\|^2 = \|R^{-1}(p^0)f(Q)\|^2$ over Q subject to the condition that Q be positive definite with least eigenvalue 1.

An optimal choice yields the largest sphere radius for the stable polynomials in the corollary.

We now present an algorithm for solving the basic Problem (A) based on Theorem 1. Denote by H the set of all stable polynomials. Then Problem (A) may be formulated thus:

$$\text{minimize } \|p(u) - h\|^2 \text{ subject to } u \in U \text{ and } h \in H. \tag{63}$$

Choose initial values $u^0 \in U$ and $h^0 \in H$. Find $p^0 = p(u^0)$. If $p^0 \in H$, the problem is solved. Otherwise construct, using Theorem 1, an ellipsoid with center at h^0 entirely contained in H; then find the point h^1 on the surface of the ellipsoid, closest to p^0.

Next, we find $p^1 = p(u^1)$ from the solution of the problem

$$\min_{u \in U} \|p(u) - h^1\|^2, \tag{64}$$

Problem (64) is solved simply if the functions $p_i(u)$ are linear. In the general case one constructs a minimizing sequence for (64), and after a suitable number of iterations finds $u^1 \in U$ and $p^1 = p(u^1)$. If $p^1 \in H$, the problem is solved; otherwise one finds the point h^2, on the surface of an ellipsoid centered at h^1, closest to p^1, and repeats.

We note that the described procedure is a relaxation process: after each cycle of minimization over u and h the value of the goal function decreases. In minimizing over h the following programming problem arises.

Given an ellipsoid $D = \{v : (v, Bv) \leq \lambda^2\}$ (cf. condition $(**)$, with $\lambda = \lambda_{\min}(Q)$) and a point w not in D, find the point $v \in D$ closest to w, i.e., find

$$\min \|v - w\|^2 \quad \text{subject to } (v, Bv) \leq \lambda^2. \tag{65}$$

Problem (65) is easily solved using Lagrange multipliers. We omit comments and only describe the algorithm:

$1°$. Find the following:

$$N = \|w\| \cdot \|y\|, \quad S = (w, y), \quad z_1 = \frac{1}{4}\left[\frac{2N}{\lambda} - 1\right], \quad z_2 = \frac{1}{2}\left[\frac{\sqrt{2N}}{\lambda} - 1\right],$$

$$a = \max\{0, \min\{z_1, z_2\}\}, \quad b = \max\{z_1, z_2\}.$$

$2°$. Construct the fourth-degree polynomial

$$f(z) = \lambda^2(1 + 2z)^2(1 + 4z)^2 - 2[N^2(1 + 4z)^2 + S^2(1 - 8z^2)].$$

$3°$. Find the unique positive root z^0 of the equation $f(z) = 0$ (it can be shown that $z^0 \in [a, b]$).

$4°$. Construct the matrix $C^{-1} = \alpha_1 I + \beta_1 yy^T$, where

$$\alpha_1 = 1/(1 + 2z^0), \quad \beta_1 = -2z^0/(\|y\|^2(1 + 2z^0)(1 + 4z^0)).$$

Then the solution of (65) is determined by $v^* = C^{-1}w$.

BIBLIOGRAPHY

1. V. N. Makeev, *The synthesis of stability of systems by the method of nonlinear programming*, Prikl. Mat. Mekh. **35** (1971), 348–351; English transl. in J. Appl. Math. Mech. **35** (1971).

2. W. Wilhelmi, *Ein Algorithmus zur Lösung eines inversen Eigenwertproblems*, Z. Angew. Math. Mech. **54** (1974), 53–55.

3. J. H. Wilkinson, *The algebraic eigenvalue problem*, Clarendon Press, Oxford, 1965.

4. B. T. Polyak, *A method of conjugate gradients in extremum problems*, Zh. Vychisl. Mat. i Mat. Fiz. **9** (1969), 807–821; English transl. in USSR Comput. Math. and Math. Phys. **9** (1969).

5. I. I. Dikin, *Iterative solution of problems of linear and quadratic programming*, Dokl. Akad. Nauk SSSR **174** (1967), 747–748; English transl. in Soviet Math. Dokl. **8** (1967).

6. Yu. M. Ermol'ev, *The method of generalized stochastic gradients and stochastic quasi-Fejér sequences*, Kibernetika (Kiev) **1969**, no. 2, 73–83; English transl. in Cybernetics **5** (1969).

7. a)* J. E. Kelley, Jr., *The cutting-plane method for solving convex programs*, J. Soc. Indust. Appl. Math. **8** (1960), 703–712.

   b) Arthur F. Veinott, Jr., *The supporting hyperplane method for unimodal programming*, Oper. Res. **15** (1967), 147–152.

8. V. P. Bulatov, *Imbedding methods in optimization problems*, "Nauka", Novosibirsk, 1977. (Russian)

9. I. S. Pace and S. Barnett, *Comparison of numerical methods for solving Liapunov matrix equations*, Internat. J. Control (1) **15** (1972), 907–915.

10. P. G. Smith, *Numerical solution of the matrix equation $AX + XA^T + B = 0$*, IEEE Trans. Automatic Control **AC-16** (1971), 278–279.

11. James Lucien Howland and John Albert Senez, *A constructive method for the solution of the stability problem*, Numer. Math. **16** (1970/71), 1–7.

12.** J. A. Senez, *On the solution of the matrix equations*, J. Math. Anal. Appl. **4** (1962), 541–550.

13. E. G. Antsiferov, *A stabiity problem for a system of linear differential equations*, Works on Applied Mathematics and Cybernetics, Sibirsk. Ènerget. Inst. Akad. Nauk SSSR, Irkutsk, 1972=Manuscript No. 5285-72, deposited at VINITI, 1972, pp. 3–6. (Russian) R. Zh. Mat. **1973**, 6Б903.

14. ____, *Application of mathematical programming methods in the problem of synthesis of statistically stable electro-energetic systems*, Application of the Method of Lyapunov Functions in Energy Research (M. A. Tagirov, editor), "Nauka", Novosibirsk, 1975, pp. 81–83. (Russian)

15. ____, *Some optimization problems connected with the construction of stable linear dynamical systems*, Applied Mathematics (V. P. Bulatov and A. P. Merenkov, editors), "Nauka", Novosibirsk, 1978, pp. 5–35. (Russian) R. Zh. Mat. **1979**, 3Б861.

16. ____, *On a sufficient condition for the stability of polynomials*, Optimization Methods and Their Applications, Sibirsk. Ènerget. Inst. Akad. Nauk SSSR, Irkutsk, 1979, pp. 93–98. (Russian) R. Zh. Mat. **1980**, 3Б1001.

Translated by O. HAJEK

Editor's note. The Russian original combines the author and title of [7a] with the journal, volume, and page span of [7b].

**Editor's note.* This reference is taken directly from the Russian: it is defective, but the intended item may be J. A. Senez, *On the solution of the matrix equation $MA = A^T M$*, Thesis, Univ. of Ottawa, Ottawa, 1969.

Amer. Math. Soc. Transl.
(2) Vol. **133**, 1986

On Solitons with Hopf Index

YU. P. RYBAKOV

§1. Introduction

Analysis of the problem of stability of three-dimensional solitons, i.e., regular, localized solutions of the nonlinear field equations, showed that solitons can only be conditionally stable [1]. Here the additional condition imposed on the initial perturbation is usually a condition of fixing some conserved quantity such as charge. If the charge is topological, i.e., it does not change under continuous deformation of the field, then this condition is automatically satisfied for all physically admissible perturbations. Therefore, solitons equipped with topological charge are the most likely candidates for the role of stable elementary particles.

Topological charges can arise in field theory either due to nontrivial boundary conditions at spatial infinity (a degenerate vacuum) [2], [3] or due to a nontrivial topology of the field itself [4], [5]. In this paper we shall be interested only in the latter case under the assumption that the physical field parametrizes some compact manifold and vanishes as $|x| \to \infty$;

$$\varphi(t, \infty) = 0. \tag{1}$$

With consideration of (1) it is possible to compactify R^3, identifying the infinitely distant points and converting R^3 into S^3. All physical fields then break down into homotopy classes which can conveniently be described by means of the third homotopy group $\pi_3(\Phi)$, assigning to each class an integer—the value of the topological charge Q. The latter is a homotopy invariant which is identically conserved, i.e., independently of the field equations.

The two simplest possibilities for the choice of the manifold Φ are $\Phi = S^3$ and $\Phi = S^2$. In the first case the topological charge is the mapping degree

1980 *Mathematics Subject Classification* (1985 *Revision*). Primary 81E08.

Translation of Problemy Teor. Gravitatsiĭ i Èlement. Chastits. Vyp. 12 (1981), 147–154; MR **84g**:58056.

©1986 American Mathematical Society
0065-9290/86 $1.00 + $.25 per page

$Q = \deg(S^3 \to S^3)$, and in the second it is the Hopf index. Among models with topological charge of degree type the Skyrme model [6] has been most studied; for this model, in particular, the existence and stability of spherically symmetric, topological solitons have been proved [1], [7], [8]. Stability of solitons is hereby realized due to the fact that the Hamiltonian in the Skyrme model can be bounded below in terms of the topological charge.

Our purpose is to study models with Hopf index, among which the Faddeev model [9], [10] is most widely known. We first describe the Hopf index.

§2. The structure of the Hopf index

We consider a triplet of scalar fields $n_a(t, \mathbf{x})$: $R \times R^3 \to S^2$, $a = 1, 2, 3$, satisfying the conditions $n_a^2 = 1$ and $n_a(t, \infty) = \delta_{a3}$, and we assign to it a 4-vector a_μ, setting

$$f_{\mu\nu} = \partial_\mu a_\nu - \partial_\nu a_\mu = 2\varepsilon^{abc}\partial_\mu n_a \partial_\nu n_b n_c. \tag{2}$$

The Hopf index is the number of mergings of the vector lines of the **b**-field where $\mathbf{b} = \operatorname{curl} \mathbf{a}$. It is not hard to see that to each point on S^2 there corresponds a closed **b**-line in R^3. We choose two points on S^2 to which there correspond two distinct **b**-lines, and we suppose that $\mathbf{b}_1$ and $\mathbf{b}_2$ merge Q times. This means that if we stretch a regularly oriented surface S on $\mathbf{b}_1$, then the line $\mathbf{b}_2$ intersects S exactly Q times. Now since each **b**-line is mapped into a point on S^2, it can be asserted that S^2 is also mapped onto S Q times. Therefore, introducing polar coordinates (β, γ) of the vector n_a, we have

$$4\pi Q = Q \int_{S^2} \sin\beta \, d\beta \, d\gamma = \int_S \sin\beta([\nabla\beta\nabla\gamma] \, d\mathbf{S}) = \frac{1}{2}\int_S (\mathbf{b}d\mathbf{S}) = \frac{1}{2}\int_{\mathbf{b}_1}(\mathbf{a}d\mathbf{l}). \tag{3}$$

Now the value of Q does not depend on the choice of the **b**-line, and we can thus write

$$4\pi Q = \frac{1}{2}\int_{\mathbf{b}_2}(\mathbf{a}d\mathbf{l}) = \frac{1}{2}\int_{\mathbf{b}}(\mathbf{a}d\mathbf{l}). \tag{4}$$

Multiplying (3) and (4), we find that

$$(4\pi Q)^2 = \frac{1}{4}\int_S (\mathbf{b}d\mathbf{S})\int_{\mathbf{b}}(\mathbf{a}d\mathbf{l}) = \frac{1}{4}\int_{S\times\mathbf{b}}(\mathbf{ab})(d\mathbf{l}d\mathbf{S}) \equiv -\frac{1}{4}Q\int_{R^3}(\mathbf{ab})d^3x.$$

Finally, we have

$$Q = -(8\pi)^{-2}\int (\mathbf{ab})d^3x = \int \mathfrak{J}^0 d^3x, \tag{5}$$

where we have introduced the identically conserved topological current

$$\mathfrak{J}^\mu = -2^{-1}(8\pi)^{-2}\varepsilon^{\mu\nu\sigma\tau}f_{\nu\sigma}a_\tau. \tag{6}$$

We now indicate an important special case where the Hopf index can be computed explicitly. This is the case of an n-field invariant under the group

$$G = \operatorname{diag}[O(2)_I \otimes O(2)_S], \tag{7}$$

where $O(2)_I$ is the group of internal 3-rotations about the axis n_3 and $O(2)_S$ is the group of spatial rotations about the Z axis. In spherical coordinates (r, ϑ, α) the invariant field has the form

$$n_3 = \cos \beta = w(r, \vartheta); \qquad \gamma = N\alpha + v(r, \vartheta), \quad N \in Z. \tag{8}$$

Noting that the vector $\mathbf{a}$ in (5) may be considered solenoidal, we use the Biot-Savart-Laplace law, with consideration of which (5) takes the form

$$Q = 2(8\pi)^{-3} \iint d^3x\, d^3x' (\mathbf{b}'[\nabla R^{-1}, \mathbf{b}]), \tag{9}$$

where $R = |\mathbf{x} - \mathbf{x}'|$ and $\mathbf{b} = -2[\nabla w \nabla \gamma]$.

By means of (8) we find

$$\mathbf{b} = -(2N/r^2 \sin \vartheta)(\mathbf{e}_r w_\vartheta - \mathbf{e}_\vartheta r w_r) - 2\mathbf{e}_\alpha r^{-1} \Im, \tag{10}$$

where $\Im = w_r v_\vartheta - w_\vartheta v_r$. Since $w \to 1$ as $r \to \infty$ and from the boundedness of $\mathbf{b}$ it follows that on the Z axis $\nabla w = 0$, we conclude that on the Z axis $w = 1$. Thus, the field (8) has toroidal structure, i.e., a surface $w = $ const is homeomorphic to the torus T^2.

Substituting (10) into (9), we obtain

$$Q = 2N(8\pi)^{-2} \int d^3x \int_0^\infty dr' \int_0^\pi d\vartheta' \Im(w'_{\vartheta'} A + w'_{r'} B), \tag{11}$$

where

$$A = (rr')^{-1}[(R^{-1})_\vartheta - \cos \alpha (R^{-1})_{\vartheta'} - \cos \alpha \cot \vartheta' R^{-1}];$$
$$B = (R^{-1})_r[\cos \alpha + (r'/r)(\cos \chi)_{\vartheta \vartheta'}] - r^{-2}(\cos \chi)_{\vartheta'}(r'/R)_\vartheta,$$
$$\cos \chi = \cos \vartheta \cos \vartheta' + \sin \vartheta \sin \vartheta' \cos \alpha.$$

In (11) it is possible to integrate by parts, taking $\nabla'_{w'}$ into $1 - w'$ with consideration of the vanishing of the function $1 - w$ on the Z axis and as $r \to \infty$. After simple transformations the integrand reduces to singular terms containing $\Delta(R^{-1}) = -4\pi\delta(\mathbf{x} - \mathbf{x}')$. As a result we obtain

$$Q = \frac{N}{4\pi} \int_0^\infty dr \int_0^\pi d\vartheta(1 - w)(w_r v_\vartheta - w_\vartheta v_r). \tag{12}$$

It is convenient to make further simplifications in cylindrical coordinates (ρ, z, α). From the symmetry of the problem it is clear that $w(\rho, z) = w(\rho, -z)$, and hence according to (12) $Q \neq 0$ only for $v(\rho, z) = -v(\rho, -z)$. Now from (8) it follows that $v(\rho, z)$ is a multivalued function with values $-n\pi/2 \leq v \leq n\pi/2$ which can have a discontinuity $[v] = \varepsilon n\pi$, $\varepsilon = \pm 1$, $n \in Z$.

Denoting the line of discontinuity by $C(\rho, z)$ and noting that the integrand in (12) is a component of the rotation, we transform the surface integral (12) into a contour integral according to Stokes' theorem. Considering the boundary conditions, we obtain

$$Q = \frac{\varepsilon}{4} nN \int_{C(\rho, z \geq 0)} (d\mathbf{l}\nabla)(1 - w)^2. \tag{13}$$

At the point $C_0 = C(\rho_0, 0)$ where the line of discontinuity terminates, the function $v(\rho, z)$ is indeterminate, $v(\rho_0, 0) = 0, \pm n\pi/2$, and hence single-valuedness of the mapping $n_a \colon R^3 \to S^2$ requires that this point correspond to a pole of the sphere S^2, i.e., we must have $w(\rho_0, 0) = \pm 1$. Now it follows from (13) that $Q \neq 0$ only for $w(\rho_0, 0) = -1$, i.e., the line of discontinuity must "pass" through the south pole. If the line of discontinuity is simply connected, then from (13) we deduce that

$$Q = \varepsilon n N = \pm n N. \tag{14}$$

Thus, identification of the line of discontinuity of the function v enables us to compute the Hopf index according to (13) or (14).

§3. The nonlinear σ-model on S^2

This model, proposed by L. D. Faddeev from geometric considerations, possesses a large degree of determinacy, since on S^2 by means of the n-field and its first derivative it is possible to construct only two independent $O(3)_I$ invariants: $f_{\mu\nu}^2$ and $(\partial_\mu n_a)^2$. However, in pure form this model possesses the drawback that the soliton solutions it admits decay too slowly at infinity (as solutions of the Laplace equation). Therefore, relativistic solitons can be obtained only in a model with violated $O(3)_I$ symmetry which, moreover, is already violated by the boundary condition $n_a(t, \infty) = \delta_{a3}$.

We therefore choose the Lagrangian density

$$\mathcal{L} = -(\varepsilon^2/4)f_{\mu\nu}^2 + \lambda^2(\partial_\mu n_a)^2 - m^2(1 - w), \tag{15}$$

where ε, λ, and m are constants. It is shown in [11] that the Hamiltonian in the model (15) admits a lower bound in terms of the Hopf index. However, the numerical value of the constant contained in this estimate is not indicated in [11]. A limit value of the constant can be found as follows.

Denoting by $\|\cdot\|_p$ the norm in $L_p(R^3)$, $p \in (1, \infty)$, we use the vector Hölder inequality

$$\int (\mathbf{u}\mathbf{v})d^3x \leq \|\mathbf{u}\|_p \|\mathbf{v}\|_q, \qquad p^{-1} + q^{-1} = 1; \tag{16}$$

the multiplicative vector inequality for L_p [12]

$$\|\mathbf{u}\|_{p_2} \leq \|\mathbf{u}\|_{p_1}^{1-\alpha} \|\mathbf{u}\|_{p_3}^{\alpha} \tag{17}$$

for $p_1 \leq p_2 \leq p_3$ and $\alpha = (1 - p_1/p - 2)(1 - p_1/p_3)^{-1}$; the identity for solenoidal vector fields

$$\|\mathbf{a}_x\|_2^2 \equiv \int d^3x (\partial_i a_k)^2 = \|\mathrm{curl}\, \mathbf{a}\|_2^2; \tag{18}$$

and the Sobolev inequality [13], [14]

$$\|\mathbf{a}\|_6 \leq \|\mathbf{a}_x\|_2 (16/27\pi^4)^{1/6}. \tag{19}$$

We observe first that for the Hamiltonian H of the model (15) we have

$$H > \int d^3x \left\{ \frac{\varepsilon^2}{2}\mathbf{b}^2 + \lambda^2[(\nabla\beta)^2 + \sin^2\beta(\nabla\gamma)^2] \right\}$$

$$\geq \int d^3x \left[\frac{\varepsilon^2}{2}\mathbf{b}^2 + 2\lambda^2\sin\beta|\nabla\beta||\nabla\gamma| \right]$$

$$\geq \int d^3x \left[\frac{\varepsilon^2}{2}\mathbf{b}^2 + \lambda^2|\mathbf{b}| \right] = \frac{\varepsilon^2}{2}\|\mathbf{b}\|_2^2 + \lambda^2\|\mathbf{b}\|_1. \tag{20}$$

Considering the boundedness of H, from (20) and (17) it follows that $\mathbf{b} \in L_p(R^3)$ for $p \in [1,2]$. Further, following [11] and setting in (16) and (17) $p = 6$, $p_2 = q = 6/5$, $p_1 = 1$, and $p_3 = 2$, we find with consideration of (18)–(20) that

$$(8\pi)^2|Q| \leq \|\mathbf{a}\|_6\|\mathbf{b}\|_{6/5} \leq d\|\mathbf{a}_x\|_2\|\mathbf{b}\|_{6/5}$$

$$\leq d\|\mathbf{b}\|_2^{4/3}\|\mathbf{b}\|_1^{2/3} < H^{4/3}(\sqrt{2}\varepsilon\lambda)^{-4/3}d,$$

where $d = (16/27\pi^4)^{1/6}$. Finally, we obtain

$$H > \varepsilon\lambda|Q|^{3/4}(4\pi)^2\sqrt{2}\,3^{3/8}. \tag{21}$$

On the static invariant fields (8) for $N = 1$ the Hamiltonian takes the form

$$H = \int d^3x\{2\varepsilon^2\sin^2\beta[\rho^{-2}(\nabla\beta)^2 + [\nabla\beta\nabla v]^2] + \lambda^2(\nabla\beta)^2$$

$$+ \lambda^2\sin^2\beta[\rho^{-2} + (\nabla v)^2] + 2m^2\sin^2(\beta/2)\}. \tag{22}$$

To determine the structure of fields minimizing the functional (22), we use the fact that the surface $\beta = \text{const}$ is homeomorphic to the torus T^2. Although for small values of the charge $Q = 1, 2, \ldots$ the angular structure of solutions is very complex, for $Q = n \gg 1$ it simplifies considerably, since the toroid becomes a closed string of large radius. In connection with this we first consider an infinite string for which

$$\beta = \Theta(\rho); \quad v = kz; \quad k = \text{const}.$$

Setting $\rho\lambda/\varepsilon\sqrt{2} = s$, $2k^2\varepsilon^2/\lambda^2 = c^2$, and $2m^2\varepsilon^2/\lambda^4 = \nu$, we reduce the problem to that of minimizing the functional

$$I[\Theta] = \int_0^\infty \frac{ds}{s}\{(\Theta')^2[s^2 + \sin^2\Theta(1 + c^2s^2)]$$

$$+ \sin^2\Theta(1 + c^2s^2) + 2\nu s^2\sin^2(\Theta/2)\}. \tag{23}$$

Here the energy found on a segment of the string of length l is equal to $H = 2\pi l\lambda^2 I[\Theta]$. The existence of string solutions can be proved by a variational method if we seek solutions satisfying the boundary conditions

$$\Theta(0) = \pi, \qquad \Theta(\infty) = 0. \tag{24}$$

Using inequalities of the type $(\Theta')^2s^2 + \sin^2\theta \geq 2s|\sin\Theta\Theta'|$ with consideration of (24) it is not hard to obtain the estimate

$$I[\Theta] \geq (\pi^2c^2 + 16)^{1/2} + (\pi^2c^2 + 128\nu/9)^{1/2},$$

which implies the existence of a minimizing sequence $\{\Theta_n(s)\}$. Moreover, from the boundedness of the integral

$$\int_d^\infty \frac{ds}{s}(y')^2 < \infty,$$

where $y = \cos\Theta$, and the Schwarz inequality, we deduce that

$$|y_2 - y_1| = \left|\int_{s_1}^{s_2} y'\, ds\right| \le \int_{s_1}^{s_2} |y'|\, ds \le \left[\int_{s_1}^{s_2}(y')^2\frac{ds}{s}\right]^{1/2}\left(\int_{s_1}^{s_2} s\, ds\right)^{1/2}$$
$$\le C|s_2^2 - s_1^2|^{1/2},$$

i.e., the function $\cos\Theta(s)$ is Hölder continuous with exponent $1/2$ on any finite interval. It may therefore be assumed that $\Theta_n \in \mathcal{H}_\Theta^1$, where $\mathcal{H}_\Theta^1$ is the Hilbert space with norm

$$\|\Theta_n\| = \left[\int_0^\infty (\Theta_n')^2 s\, ds + \Theta_n^2(a)\right]^{1/2}, \qquad a \in (0,\infty).$$

Thus, the sequence $\{\Theta_n\}$ converges weakly in $\mathcal{H}_\Theta^1$ to a function $\Theta(s)$. Applying further the scheme of proof used in [1] and [8], it is not hard to establish that the functional (23) attains its infimum, and also the regularity of the function $\Theta(s)$ which has the following asymptotic behavior:

$$s \ll 1\colon \Theta \approx \pi - as;$$
$$s \gg 1\colon \Theta \approx bs^{-1/2}\exp[-s(2c^2 + \nu)^{1/2}].$$

We now close the string. We note that the vector lines of the **b**-field are helical lines with lead $2\pi/k$. Therefore, in closing a segment of the string of length l on which are packed $kl/2\pi = n$ leads of the helical line, the number of mergings of **b**-lines is equal to n, whence $Q = n$ and $l = 2\pi n/k = 2\pi\varepsilon\sqrt{2}n/\lambda c$. As a result, we find the energy of the closed string as a function of the parameter c:

$$H(c) = \varepsilon\lambda n 4\pi^2\sqrt{2}I(c,\nu)/c. \tag{25}$$

Numerical calculations carried out by Kundu indicate that the function $I(c,1)/c$ has a minimum for $c \approx 1.9$ equal to 10.6, which when substituted into (25) agrees well with the estimate (21).

§4. On spinor realization of the σ-model

If the topological charge is interpreted as a baryon, then in the relativistic model topological solitons must have half-integer spin. The simplest means of providing this is to include fermion fields. We proceed from the internal group

$$G = SU(2) \otimes U(1)_e \otimes U(1)_l, \tag{26}$$

whose structure can, for example, be connected with the laws of conservation of isospin and electric and lepton charges. We identify the group $U(1)_l$ with the group $O(2)_I$. This correspondence is not accidental, since it provides, as we shall soon see, nonintersectability of the classes of leptons and baryons.

We introduce a spinor field ψ which is an $SU(2)$-doublet and establish its correspondence with the n-field:

$$n_1 = \overline{\psi}\psi; \qquad n_2 = i\overline{\psi}\gamma^5\psi. \tag{27}$$

The groups $U(1)_e$ and $U(1)_l$ are then realized by the transformations

$$\psi \to \exp(i\alpha)\psi; \qquad \psi \to \exp(i\alpha\gamma^5)\psi,$$

where γ^5 can be chosen to be diagonal. Our aim is to form spinor invariants bounding the Hamiltonian of the model (15) above and not vanishing in the lepton limit where $\gamma^5\psi = \pm\psi$. In this case, according to (27), $n_1 = n_2 = 0$, which automatically leads to the vanishing of the baryon charge—the Hopf index.

We begin with invariants without derivatives. The requirement of invariance relative to the group (26) picks out the vectors $\overline{\psi}\gamma^\mu\psi$, $\overline{\psi}\gamma^5\gamma^\mu\psi$, $\overline{\psi}\gamma^\mu\tau\psi$, and $\overline{\psi}\gamma^5\tau\psi$. However, only the first of these has a square of definite sign: $j^\mu = \overline{\psi}\gamma^\mu\psi$. We have

$$j^2 \equiv j_\mu^2 \geq n_1^2 + n_2^2, \tag{28}$$

for the proof of which we set $\psi = \psi_1 \oplus \psi_2$, where $2\psi_{1,2} = (1 \pm \tau_3)\psi$. We write the form

$$2\Phi = (\overline{\psi}\gamma^\mu\psi)^2 - (\overline{\psi}\psi)^2 - (i\overline{\psi}\gamma^5\psi)^2$$

and set

$$j_a^\mu = \overline{\psi}_a\gamma^\mu\psi_a; \quad s_a = \overline{\psi}_a\psi_a; \quad p_a = i\psi_a\gamma^5\psi_a, \qquad a = 1, 2.$$

With consideration of the Fierz identities $j_a^2 = s_a^2 + p_a^2$ the form Φ can be written as

$$\Phi = (j_1 j_2) - (s_1 s_2 + p_1 p_2).$$

Choosing a reference system in which $j_2 = 0$ and setting $s_2 = j_2^0 \cos\alpha$ and $p_2 = j_2^0 \sin\alpha$, we represent Φ as a quadratic form in ψ_1:

$$\Phi = j_2^0 \psi_1^+ [1 - \gamma^0 \exp(i\alpha\gamma^5)]\psi_1,$$

which is obviously nonnegative, since the eigenvalues of the matrix $\gamma^0 \exp(i\alpha\gamma^5)$ are ± 1.

Inequality (28) makes it possible in (15) to replace the term $1 - w$, which vanishes in the lepton limit, by $1 - (1 - j^2)^{1/2} \geq 1 - w$. As concerns terms with derivatives, we need invariants which give positive terms in the Hamiltonian. In forming them we use the nonnegative matrix $\gamma^0\gamma^\mu j_\mu = B$. The validity of the inequality

$$4\dot{\psi} + B\dot{\psi} \geq \dot{n}_1^2 + \dot{n}_2^2, \tag{29}$$

where $\dot{\psi} = \partial_0\psi$, can be established by direct verification.

Thus, on the basis of (28) and (29) it is possible to propose the following Lagrangian density which satisfies the conditions enumerated above:

$$\mathcal{L} = -(\varepsilon^2/4)f_{\mu\nu}^2 + \lambda^2(\partial_\mu w)^2 + \sigma^2 j^\mu(\partial_\nu\overline{\psi}\gamma_\mu\partial^\nu\psi) - m^2[1 - (1 - j^2)^{1/2}], \tag{30}$$

where $j^\mu = \overline{\psi}\gamma^\mu\psi$. For $\sigma^2 \geq 4\lambda^2$ the Hamiltonian of the model (30) satisfies (21) and hence admits the existence of stable topological solitons. We note that if we introduce in (30) terms linear in the derivatives, then (21) will be violated globally, but for some selected states (with restrictions on $|\psi|$) (21) may still be satisfied.

BIBLIOGRAPHY

1. Yu. P. Rybakov, *Conditional stability of regular solutions in a nonlinear field theory*, Problemy Teor. Gravitatsii i Èlementar. Chastits Vyp. 10 (1979), 194–202. (Russian)

2. Luis J. Boya, José F. Cariñena, and Juán Mateos, *Homotopy and solitons*, Fortschr. Phys. **26** (1978), 175–214.

3. B. Felsager, *Geometry of solitons*, Copenhagen, 1977.

4. L. D. Faddeev, *In search of multidimensional solitons*, Nonlocal, Nonlinear and Nonrenormalizable Field Theories (Materials Fourth Internat. Congr., Alushta, 1976; D. I. Blokhintsev, editor), Ob″ed. Inst. Yadern. Issled., Dubna, 1976, pp. 207–223. (Russian)

5. ____, *Some comments on the many-dimensional solitons*, Lett. Math. Phys. **1** (1975/77), 289–293.

6. T. H. R. Skyrme, *A unified field theory of mesons and baryons*, Nuclear Phys. **31** (1962), 556–569.

7. V. N. Romanov, I. V. Frolov, and A. S. Shvarts, *Spherically symmetric solitons*, Teoret. i Mat. Fiz. **37** (1978), 305–318; English transl. in Theoret. Math. Phys. **37** (1978).

8. A. Kundu, Yu. P. Rybakov, and V. I. Sanyuk, *Topological solitons in the Skyrme gauge model*, Indian J. Pure Appl. Phys. **17** (1979), 673–677.

9. L. D. Faddeev, *Gauge-invariant model of electromagnetic and weak lepton interactions*, Dokl. Akad. Nauk SSSR **210** (1973), 807–810; English transl. in Soviet Phys. Dokl. **18** (1973/74).

10. ____, Preprint MPI-PAE/Pth 16, Max-Planck-Inst., 1974.

11. A. F. Vakulenko and L. V. Kapitanskiĭ, *Stability of solitons in S^2 of a nonlinear σ-model*, Dokl. Akad. Nauk SSSR **246** (1979), 840–842; English transl. in Soviet Phys. Dokl. **24** (1979).

12. O. A. Ladyzhenskaya, *Boundary value problems of mathematical physics*, "Nauka", Moscow, 1973; English transl., Springer-Verlag, 1985.

13. S. L. Sobolev, *Applications of functional analysis in mathematical physics*, Izdat. Leningrad. Gos. Univ., Leningrad, 1950; reprint, Izdat. Sibirsk. Otdel. Akad. Nauk SSSR, Novosibirsk, 1962; English transl., Amer. Math. Soc., Providence, R.I., 1963.

14. Gerald Rosen, *Minimum value for c in the Sobolev inequality* $\|\phi^3\| \leq c\|\nabla\phi\|^3$, SIAM J. Appl. Math. **21** (1971), 30–32.

Translated by J. R. SCHULENBERGER

Amer. Math. Soc. Transl.
(2) Vol. **133**, 1986

Complete Integrability of Some
Classical Hamiltonian Systems

UDC 517.54

A. T. FOMENKO

§1. A general view of the mechanisms of integrability

The integration of Hamiltonian systems in explicit form is far from always successful (and often the very term "explicit integration" requires nontrivial definition), and therefore the theorem of Liouville (see [1]) is a highly useful instrument for describing the solutions of Hamiltonian systems.

Many important Hamiltonian systems can be realized as Hamiltonian systems (relative to the canonical form on the orbits, described above) on corresponding Lie (co-) algebras. As the results of recent years have shown, a natural approach to the analysis of problems of integrability is the following: we fix on the Lie (co-) algebra G a symplectic structure (for example, the one described above) and look for a family $F(G)$ of smooth functions on the orbits (in the algebra) of the (co-) adjoint representation satisfying the following requirements: 1) the space of functions $F(G)$ forms a commutative Lie algebra (relative to the Poisson bracket), and 2) in the space of functions $F(G)$ an additive basis $f_1, \ldots, f_k$ of functionally independent functions can be chosen, where $k = \frac{1}{2} \dim \mathcal{O}$, i.e. there are sufficiently many of them for Liouville's theorem to be applicable. In other words, we seek a family of functionally independent functions, equal in number to half the dimension of the orbit, it being necessary in the first place to solve the problem for orbits of maximal dimension (generic orbits). In the case of a compact semisimple group the desired family of functions has to have dimension $\frac{1}{2}(n - r)$, where n is the dimension of the algebra G and r is the rank of G (i.e. the dimension of a maximal commutative subalgebra, a so-called Cartan subalgebra). For this it is natural to look straightaway for

1980 *Mathematics Subject Classification* (1985 *Revision*). Primary 58F05, 58F07; Secondary 17B99.

Translation of Monogenic Functions and Mappings (N. P. Korneĭchuk, editor), Inst. Mat. Akad. Nauk Ukrain. SSR, Kiev, 1982, pp. 3–19; MR **84k**:58115.

©1986 American Mathematical Society
0065-9290/86 $1.00 + $.25 per page

functions $f_1, \ldots, f_k$ defined on the whole Lie algebra G, whose restrictions to generic orbits give the desired commutative family of maximal dimension. In the semisimple case, for example, it is possible (see [2] and [3]) to find a collection of $\frac{1}{2}(n-r)+r$ independent functions, commutative relative to the Poisson bracket, defined on the whole algebra G (and not only on orbits). Here r functions of this collection are invariants of the adjoint representation (i.e. their common level surface defines a generic orbit) while the remaining $\frac{1}{2}(n-r)$ functions are nonconstant on the orbits. Recall that the symplectic form described above is, generally speaking, degenerate on the whole algebra (as distinct from the orbits, on which it is nondegenerate).

Clearly, if for some Lie algebra G we succeed in constructing a collection of functions $F(G)$ (commutative and maximal) then, taking any one of the functions of this collection as a Hamiltonian, we can consider the Hamiltonian vector field sgrad f, completely Liouville integrable on orbits since all the remaining functions of the collection $F(G)$ are first integrals of this Hamiltonian system.

In this approach, the basic problem is the construction of as large as possible a supply of commutative and maximal families of functions on the orbits of the (co-) adjoint representation. The larger this supply, the more examples we obtain of completely Liouville integrable Hamiltonian systems. Along with this approach there naturally arises a converse problem: given some concrete system whose complete integrability we wish to obtain, for example, in terms of Liouville's theorem, we have to try to represent this system in the form of a Hamiltonian vector field on an orbit of the (co-) adjoint representation of some Lie (co-) algebra or other. Here, as concrete examples show (see, for example, [2] or [4]–[6]), the same systems can appear as Hamilton fields on orbits of different Lie (co-) algebras, these realizations generally speaking having substantially different algebraic properties. By varying the Lie (co-) algebra we can try to discover a representation of the given system on which it will have the form sgrad f, where $f \in F(G)$ for some commutative and maximal family of functions on the Lie algebra G. This also gives a full collection of integrals of the system. Sometimes it is possible to algebraicize the given system, i.e. represent it in the form of a field sgrad f (on an underlying Lie (co-) algebra) not on all generic orbits, but only on one orbit or another (not excluding, in particular, that the orbit may be singular). Then, the system being algebraicized (if this is possible), there comes the next stage—the search for a complete supply of integrals (i.e. a maximal commutative family of independent functions on the algebra or the orbit). At present there are a few methods that allow integrals to be constructed on the orbits of a coadjoint representation. The method proposed for finite-dimensional Lie algebras in [2], based on the idea of translation of invariants of the adjoint representation on generic covectors, has proved to be highly effective. The working out of this idea has led to the appearance of methods of constructing integrals by means of translation (by a generic covector) of semi-invariants of a Lie (co-) algebra, i.e. functions which are multiplied by the character of the (co-) adjoint representation under the (co-) adjoint action of the group: see [7]–[9]. In [8] and [9] methods of constructing integrals (of maximal commutative families of functions) are worked out, based on the algebraic properties of a chain of Lie subalgebras in the Lie algebra under investigation.

As the experimental material amassed up to the present has shown, an important place among Liouville integrable systems is held by systems related to left-invariant metrics of the form $\varphi_{a,b,D}$ introduced in [2]. Although up to now no invariant metric characteristic of these metrics has been discovered, distinguishing them in the class of all left-invariant metrics (on semisimple groups, for example), nonetheless, from the fact of complete integrability of the corresponding equations of motion (where these metrics serve as Hamiltonians) it should follow that the common level surfaces of the integrals possess some kind of homogeneity properties, not excluding for example that some of those surfaces (tori of half the dimension in orbits of the coadjoint representation) may be minimal submanifolds or even completely geodesic with respect to the metric $\varphi_{a,b,D}$. If the group (algebra) is not semisimple, then the operators participating in the formulation of the metric $\varphi_{a,b,D}$ (see [2]) are absent. Therefore in the case of an arbitrary finite-dimensional Lie algebra the question of the existence of a maximal commutative collection of functions is not yet solved (although most probably this question will be answered in the positive sense). This problem is connected also with the existing reduction of the noncommutative Liouville theorem (proved in [3]) to the commutative Liouville theorem; in other words, if the hypothesis on the existence of a maximal commutative collection of functions on an arbitrary finite-dimensional Lie algebra is valid, the presence of a maximal noncommutative algebra of integrals of a Hamilton system on a symplectic manifold automatically entails the existence also of a maximal commutative algebra of integrals.

Here we shall give our main consideration to a survey of results recently obtained in the direction of solving the problems of complete integrability of new, interesting Hamiltonian systems connected with Lie algebras (not necessarily compact). In particular, we shall describe (apart from results of the author expounded in [1]) some results of his students obtained in 1980–1981. Many of these results were obtained at the author's research seminar in the Mechanics-Mathematics Faculty of Moscow State University. For example, we have succeeded in integrating (Liouville) the equations of inertial motion of a many-dimensional rigid body in an ideal fluid (where in the particular case of three-dimensional Euclidean space cases of integration of classical systems with indefinite energy are obtained, these new cases of integrability being at the same time analogous in algebraic structure to the first case of Clebsch), analogs of these equations on the Lie algebras $u(p,q) \times \mathbf{C}^{p+q}$, the equations of magnetohydrodynamics, and so on. Many of these results were obtained thanks to a significant clarification of the algebraic structure of integrable Hamilton systems on Lie algebras. Here a definitive role was played by the general construction of section operators, proposed by the author for an arbitrary linear representation with the aim of distinguishing a special class of systems that naturally includes basic cases of complete integrability. Recall that the idea of this construction reduces briefly to the following:

Let H be a Lie algebra with $\mathfrak{H}$ the corresponding Lie group, let $\rho\colon H \to \mathrm{End}\,V$ be a presentation of H on the linear space V, let $\alpha = \exp(\rho)$ be the corresponding representation of the group, and let $\mathcal{O}(X)$ denote orbits of the action of $\mathfrak{H}$ on V, where $X \in V$. If we are given a linear operator (which we call a *section operator*) $C\colon V \to H$, then there arises a natural vector field $\dot{X}_C = \rho(CX)X$ on the orbits. The provision of a section operator also sometimes allows a symplectic structure to be defined on an orbit. For applications an important role is played by a special class of section operators (see their description above) that form a multiparameter family, whose basic parameters are a pair of elements $a \in V$, $b \in \ker\Phi_a$, where $\Phi_a(h) = (\rho h)a$. For example, in the case where $H = \mathrm{so}(n)$ and $\rho = \mathrm{ad}$ the field $\dot{X}_C$ coincides with the equations of motion of a many-dimensional rigid body with a fixed point (in the absence of gravity). An important property of this multi-parameter family of section operators is that it allows us to unify, from a common point of view, the construction of many-dimensional analogs of systems connected with the motion of a rigid body (and admitting complete integrability). Indeed this unification reveals a general algebraic mechanism for the appearance of integrals in systems of this type.

In particular when $G = H + V$ coincides with the standard splitting of a compact symmetric space (H is the stationary subalgebra), the field $\dot{X}_C$ becomes the flow on an orbit of the action of the stationary subgroup. For the case of a sphere, a complete classification of all Hamiltonian flows for the form $\dot{X}_C$ has been obtained (see [1]). Necessary and sufficient conditions have been found ensuring the existence on an orbit of symplectic structures which are invariant with respect to the field $\dot{X}_C$. This allows us to study multidimensional analogs of the equations of motion of a "singular rigid body," i.e., when the matrix of moments of inertia satisfies a system of linear relations. For the case of the group of motions of $\mathbf{R}^n$ the system $\dot{X}_C$ becomes (for the special series of section operators see [1]) the equations of motion of an inertial rigid body in an ideal fluid admitting, as it turned out, a complete collection of commuting integrals on an orbit of the coadjoint representation of this group (see [5]). Polynomial integrals are produced explicitly.

From a series of results recently obtained, we pick out the construction by V. V. Trofimov which, starting from a Lie algebra with a complete (maximal) commutative collection of functions, allows us to construct a family of new Lie algebras (which are not isomorphic to the original one, generally speaking) also possessing a maximal commutative collection of functions. Recently A. V. Brailov worked out a technique for finding a maximal collection of commuting functions on some Lie algebras appearing as "contractions" of larger Lie algebras. This has allowed equations of type $\dot{X}_C$ to be completely integrated (Liouville) on all the Lie algebras $(u(p,q) \times \mathbf{C}^{p+q})^*$, where $u(p,q) \times \mathbf{C}^{p+q}$ is the semidirect product of the Lie algebra $u(p,q)$ with its standard representation of minimal dimension. Exploiting Trofimov's results, Brailov succeeded in constructing a maximal collection of commuting functions on Lie algebras which are tensor products of Lie algebras already possessing a complete such collection (for example, semisimple ones), and on algebras equipped with Poincaré duality. This adds new arguments in favor of the conjecture on the existence of a complete collection of commuting functions on any finite-dimensional Lie algebra.

§2. The complete integration of the equations of inertial motion of a many-dimensional rigid body in an ideal fluid

These results have been announced in [5]. The Lie algebra $E(n)$ of the group of motions of Euclidean space is the semidirect sum $so(n) \oplus_\varphi \mathbf{R}^n$, where $\varphi\colon so(n) \to \mathrm{End}(\mathbf{R}^n)$ is the derivative of the standard representation of minimal dimension of the group $SO(n)$ on $\mathbf{R}^n$. The Lie algebra $E(n)$ admits a matrix realization

$$\left(\begin{array}{ccc|c} & & & y_1 \\ & so(n) & & \vdots \\ & & & y_n \\ \hline 0 & \cdots & 0 & 0 \end{array} \right),$$

the space $E(n)^*$ having the form

$$\left(\begin{array}{ccc|c} & & & 0 \\ & so(n) & & \vdots \\ & & & 0 \\ \hline y_1 & \cdots & y_n & 0 \end{array} \right).$$

The minor of the matrix X positioned at the intersection of the rows with numbers $i_1, \ldots, i_s$ and columns with numbers $j_1, \ldots, j_s$ is denoted by $M^{i_1 \cdots i_s}_{j_1 \cdots j_s}(X)$. We define functions $F^{i_1 \cdots i_s}_{j_1 \cdots j_s}(X)$ on $E(n)$ and $E(n)^*$ by the equation

$$F^{i_1 \cdots i_s}_{j_1 \cdots j_s}(X) = M^{i_1 \cdots i_s}_{j_1 \cdots j_s}(\tfrac{1}{2}(X - X^T))$$

where T is the operation of transposition relative to the main diagonal.

PROPOSITION 2.1. *The functions $J_k(X)$ on $E(n)^*$, where*

$$J_k(X) = \sum_{1 \le i_1 < \cdots < i_k \le n} F^{i_1 \cdots i_k, n+1}_{i_1 \cdots i_k, n+1}(X),$$

are invariants of the coadjoint representation of the group of motions of Euclidean space.

PROPOSITION 2.2. *Let $\mathcal{O}$ be an orbit of maximal dimension of the coadjoint representation Ad^* of the Lie group $\mathcal{E}(n)$ of motions of Euclidean n-space. Then $\mathrm{codim}\, \mathcal{O} = [(n+1)/2]$. The collection presented in Proposition 2.1 of invariants of the coadjoint representation Ad^* of the Lie group $\mathcal{E}(n)$ is a complete collection of invariants, i.e.. any other invariant can be functionally expressed in terms of the invariants J_k.*

PROPOSITION 2.3. *Let $\xi \in \mathrm{so}(n)$, $a \in \mathbf{R}^n$, $S \in \mathrm{so}(n)^* \cong \mathrm{so}(n)$, $M \in (\mathbf{R}^n)^* \cong \mathbf{R}^n$, and*

$$a + \xi \in \mathrm{so}(n) \oplus_\varphi \mathbf{R}^n = E(n),$$
$$S + M \in E(n)^* = (\mathrm{so}(n) \oplus \mathbf{R}^n)^* \cong \mathrm{so}(n)^* \oplus (\mathbf{R}^n)^* = \mathrm{so}(n) \oplus \mathbf{R}^n.$$

Then

$$\{\xi + a, S + M\}|_{\mathrm{so}(n)} = [S, \xi] + \tfrac{1}{2}(Ma^T - aM^T),$$
$$\{\xi + a, S + M\}|_{\mathbf{R}^n} = -\xi M.$$

Here M, $a \in \mathbf{R}^n$ are written as column vectors, and T is the transposition operator.

Let us now consider the section operator $C(a, b, \mathcal{D})$, constructed in the preceding paragraphs. It turns out that its corresponding Euler equations not only coincide with the equations of inertial motion of a rigid body in an ideal fluid, but also admit a complete collection of integrals. These integrals are polynomials (of various degrees).

THEOREM 2.1. *The system of differential equations $\dot{x} = \{C(a, b, \mathcal{D})(x); x\}$ on the coalgebra $E(n)^*$ is completely integrable on the generic Ad^*-orbits of the Lie group $\mathcal{E}(n)$. Let F be a function on $E(n)^*$ invariant with respect to the coadjoint representation $\mathrm{Ad}^*(\mathcal{E}(n))$. Then the functions $F_\lambda(x) = F(x + \lambda a)$ are first integrals of the motion for any numbers λ. Any two integrals $F_\lambda(x)$ and $G_\mu(x)$ are in involution on the orbits of the Ad^* representation of the Lie group $\mathcal{E}(n)$, the number of independent integrals of the given from being equal to half the dimension of the generic orbit.*

Recall the classical equations of inertial motion of a rigid body in an ideal fluid (for details see, for example, [10] or [11]). We attach a system of coordinates to the moving body; let u_i be the components of velocity of the translational motion of the origin, and w_i the components of angular velocity of rotation of the rigid body. Then the kinetic energy of the system (fluid-rigid body) had the form

$$T = \tfrac{1}{2}(A_{ij} w_i w_j + B_{ij} u_i u_j + C_{ij} w_i u_j)$$

where A_{ij}, B_{ij}, and C_{ij} are constants, depending on the forms of the body and the densities of the body and fluid; pairwise repeated indices are summed. Let $N = (y_1, y_2, y_3)$, where $y_i = \partial T / \partial w_i$; let $K = (x_1, x_2, x_3)$, where $x_i = \partial T / \partial u_i$. Then the inertial motion of the rigid body in an ideal fluid is described by the equations

$$dN/dt = N \times w + K \times U, \qquad dK/dt = K \times w,$$

where $U = (u_1, u_2, u_3)$ and $w = (w_1, w_2, w_3)$.

The kinetic energy T of the rigid body is a homogeneous quadratic form (in the classical cases it is positive definite) in the six variables u_i, w_i; it is defined by the 21 coefficients A_{ij}, B_j, C_j. These equations in the three-dimensional case have three classical integrals of Kirchhoff: $T = \text{const}$, $x_1^2 + x_2^2 + x_3^2 = \text{const}$, and

$x_1y_1 + x_2y_2 + x_3y_3 = \text{const}$. For complete integrability of the equations in the three-dimensional case we have to have four functionally independent integrals. The corresponding Hamiltonian flow is expressed by a vector field on a four-dimensional orbit. Two of the integrals "cut out" this orbit from the enveloping six-dimensional algebra, and two integrals have to be in involution on the orbit in order to foliate it into "Liouville tori" on which the motion of the system takes place. Three classical cases are well known when there is an additional fourth integral, i.e. the given equations can be integrated in explicit form (see [12]). The first general solution of the equations of motion of rigid body in a fluid was given by Kirchhoff for a solid of revolution. In 1871 Clebsch indicated two new forms of the kinetic energy, through which it was possible to find an additional fourth integral and, consequently, to bring the problem to quadrature. The solution of the problem for the first Clebsch case, when the fourth integral is generally speaking a homogeneous linear function of x_i and y_i, was proposed by Halphen (see [13] and [14]). In 1878 Weber investigated the second Clebsch case, where the fourth integral is expressed by a homogeneous quadratic function of x_i and y_i, with a particular assumption on the choice of arbitrary constants (see [15]). A third form of the kinetic energy, through which the equations can be explicitly integrated, was exposed by Steklov (see [11]). All these cases are characterized by the fact that they impose restrictions on the "configuration of the rigid body," it being supposed that the body is invariant under special symmetries of one kind or another.

The following result can be established by direct calculation.

THEOREM 2.2. *For $n = 3$ the system of differential equations*

$$\dot{x} = \{C(a, b, \mathcal{D})(x), x\}$$

on $E(n)^*$ constructed above based on a section operator coincides with the equations of inertial motion of a rigid body in an ideal fluid (in three-dimensional space).

On this basis and starting from the explicit form of the Euler equations that we constructed, we consider these equations for arbitrary dimension as describing the motion of a many-dimensional body in a fluid.

It is interesting that even in the simplest case $n = 3$ we obtain a new case of integrability, in which the kinetic energy is indefinite in sign and the additional fourth integral is linear. However, this additional integral is linear only in the case of three-dimensional space; in dimensions four and above the additional integrals are polynomials of various degrees. Computation of the kinetic energy in explicit form shows that although this energy is not positive definite, in algebraic structure this case of integrability is to a considerable extent analogous to the first Clebsch case for positive forms. The explicit form of the kinetic energy matrix (for the case of indefinite sign) corresponding to the integrable case is as follows.

$$A = \begin{pmatrix} -2\alpha & 0 & 0 & 0 & 0 & \frac{\gamma}{2} - \beta \\ 0 & 0 & 0 & 0 & \frac{-2a_1}{b_2} & 0 \\ 0 & 0 & 0 & \frac{a_1}{b_2} & \frac{a_1}{b_2} & 0 \\ 0 & 0 & \frac{a_1}{b_2} & b_2^{-2}(-b_2 a_2 + 2b_1 a_1) & 0 & 0 \\ 0 & \frac{-2a_1}{b_2} & \frac{a_1}{b_2} & 0 & b_2^{-2}(-b_2 a_2 + 2b_1 a_1) & 0 \\ \frac{\gamma}{2} - \beta & 0 & 0 & 0 & 0 & \delta \end{pmatrix}$$

Here α, β, γ, and δ are constants defining the operator $D\colon K^* = \mathbf{R}^2 \to K = \mathbf{R}^2$ that is involved in the definition of the section operator. An elementary calculation shows that the kinetic energy can be reduced to the following diagonal form:

$$T = -2\alpha \left(f_1 - \frac{\gamma - 2b}{4\alpha} u_3 \right)^2 + \left[\frac{(\gamma - 2\beta)^2}{8\alpha} + \delta \right] u_3^2 - \lambda \left(u_1 - \frac{a_1}{b_2 \lambda} f_3 \right)^2$$
$$- \lambda \left(u_2 + \frac{2a_1}{b_2} f_2 - \frac{a_1}{b_2 \lambda} f_3 \right)^2 + \frac{(2a_1)^2}{b_2 \lambda} \left(f_2 - \frac{f_3}{2b_2} \right)^2$$
$$+ \frac{2a_1^2 (b_2 - 1)^2}{b_2^3 \lambda} f_3^2.$$

The corresponding system of equations on the orbit can be explicitly integrated, as long as the additional fourth integral is linear. In higher dimensions (beginning with four) this procedure for explicit integration becomes rapidly more complicated.

§3. The construction of Lie algebras with a complete involutive collection of functions on generic orbits

As we have noted, up to now the existence of a maximal commutative collection of functions on generic orbits has been proved for many important series of Lie algebras (semisimple, Borel, and so on; see [1]). Along with this procedure for the analysis of important series of algebras encountered in diverse applications, we propose another approach to the construction of Lie algebras with a maximal commutative collection of functions. Consider any Lie algebra on which such a collection exists. How, starting from this one, can we construct a new Lie algebra with the same property? It turns out that there is a natural algorithm allowing the construction of a whole series of new such "extended" Lie algebras with a maximal commutative collection of functions. In this section we shall very briefly describe new results obtained by A. V. Brailov. Some of them rely upon the scheme proposed by V. V. Trofimov.

DEFINITION. We call a finite-dimensional graded commutative algebra over the field of real numbers
$$A = A_0 \oplus \cdots \oplus A_n, \qquad A_i \cdot A_j \subset A_{i+j} \quad (i + j \leq n)$$
an *algebra with Poincaré duality* if $\dim A_n = 1$ and the symmetric bilinear pairing $\langle \, , \, \rangle_S \colon A \times A \to \mathbf{R}$ given by the formulas
$$\langle a, b \rangle_S = 0, \quad a \in A_i,\ b \in A_j,\ i + j \neq n, \qquad \langle a, b \rangle_S a_n = ab, \quad ab \in A_n$$
is nondegenerate for a nonzero element $a_n \in A_n$.

Let $\dim A = N$. Then we choose in the algebra A a basis $B = \{\varepsilon_i\}_1^N$ of the following form: $\varepsilon_1 = +1$; then we choose bases in the subspaces A_i, proceeding from the lower grades to the higher, such that if $i < n/2$ then the bases are chosen arbitrarily, if $i = n/2$ then the basis is chosen so that the bilinear form described above is diagonal in this basis, and if $i > n/2$ then the bases are chosen conjugate to the bases of lower grades in the subspaces A_{n-i}. Instead of the field $\mathbf{R}$ we could, on the other hand, take any field k.

It turns out that these algebras are convenient for extensions of Lie algebras possessing a maximal commutative collection of functions. These extensions are constructed by the following simple method.

Let G be a Lie algebra over a field k, and let A be an algebra with Poincaré duality (over k); consider the tensor product $G_A = G \otimes A$. Clearly, this algebra becomes a Lie algebra over the field k. Here $\dim A = N$ and $\dim G = m$.

Let $P \in k[x_1, \ldots, x_m]$ be an arbitrary polynomial, regarded as a function on G^* (here in the algebra G and the coalgebra G^* we fix conjugate bases). Let $e_1, \ldots, e_m$ be the basis in G^*. Then as a basis in the new Lie coalgebra G_A^* we take the collection of elements

$$e_1^{\varepsilon_1}, \ldots, e_m^{\varepsilon_1}, \; e_1^{\varepsilon_2}, \ldots, e_m^{\varepsilon_2}, \; e_1^{\varepsilon_3}, \ldots, e_m^{\varepsilon_3}, \; \ldots, \; e_1^{\varepsilon_N}, \ldots, e_m^{\varepsilon_N},$$

where $\varepsilon_i \in B$, the basis in A.

Note. Here for convenience we have used the symbol $e_i^{\varepsilon_\alpha}$ instead of tensor product notation.

We shall denote the corresponding coordinates in the coalgebra G_A^* by $x_1^{\varepsilon_1}, \ldots, x_m^{\varepsilon_N}$. Here we assume that in the algebra A we have fixed an operator $*\colon A \to A$ of definite sign which gives rise to the bilinear pairing described above, i.e.

$$\langle *\varepsilon_i, \varepsilon_j \rangle = \delta_{ij}, \qquad *\varepsilon_i \in B, \qquad *^2 = 1.$$

Now we describe a construction which allows us to build a function (polynomial) on G_A^* from a function (polynomial) on G^*. Indeed this procedure allows us also to construct new integrals of the new (extended) algebras, knowing integrals on the original Lie algebra.

Let $P \in k[x_1, \ldots, x_m]$ be a polynomial on G^*; then we define the polynomial $\nu P \in A[x_1^{\varepsilon_1}, \ldots, x_m^{\varepsilon_N}]$ in the following way (here $\varepsilon_N = *1$):

$$\nu P(x_1^{\varepsilon_1}, \ldots, x_m^{\varepsilon_N}) = \sum_{k=0}^{\infty} \sum_{i_1 \ldots i_k} \sum_{j_1 \ldots j_k} P_{i_1 \ldots i_k} \varepsilon_{j_1 \ldots j_k} x_{i_1}^{*\varepsilon_1} \ldots x_{i_k}^{*\varepsilon_{j_k}}$$

$$= \sum_{I,J} P_I \varepsilon_J X_I^{*\varepsilon_J}$$

in conventional abbreviated notation.

By direct computation it may be verified that the mapping

$$\nu \colon k[x_1, \ldots, x_m] \to A[x_1^{\varepsilon_1}, \ldots, x_m^{\varepsilon_N}]$$

is a k-algebra homomorphism. We note that the symmetric pairing for elements of the algebra A as described above gives rise to a natural inner product which we denote by $\langle\,,\,\rangle$. In fact, as $\langle a, b\rangle_S a_n = ab$ if $ab \in A_n$, since $\dim A_n = 1$ we obtain an inner product $A \times A \to \mathbf{R}$.

Now we describe an operation, generalizing "translation by a generic covector," i.e. generalizing the basic operation introduced in [2] for arbitrary Lie algebras and, as it has turned out, leading in many situations to the discovery of a maximal commutative collection of functions.

For an element $a \in A$ we define $\nu_a P \in k[x_1^{\varepsilon_1}, \ldots, x_m^{\varepsilon_N}]$:

$$P^a(x_1^{\varepsilon_1}, \ldots, x_m^{\varepsilon_N}) = \nu_a P(x_1^{\varepsilon_1}, \ldots, x_m^{\varepsilon_N}) = \sum_{I,J} P_I \langle a, \varepsilon_J \rangle x_I^{*\varepsilon_J}.$$

THEOREM 3.1. *If the polynomials $P_1, \ldots, P_r \in k[x_1, \ldots, x_m]$ are (functionally) independent at the point $y \in G^*$, $y = (y_1, \ldots, y_m)$, then the polynomials $P_1^{\varepsilon_1}, \ldots, P_r^{\varepsilon_n} \in k[x_1^{\varepsilon_1}, \ldots, x_m^{\varepsilon_N}]$ are independent at any point $x \in G_A^*$ at which the leading coordinates have the form $x_i^{*\varepsilon_1} = y_i$, $i = 1, \ldots, m$.*

The proof goes by direct calculation.

As was shown in [2], translations of invariants are in involution. Analogous assertions hold in the present situation. More precisely, the following formula is valid: if $a, b \in A$ and $P, Q \in k[x_1, \ldots, x_m]$ then the Poisson bracket is given thus:

$$\{\nu_a P, \nu_a Q\} = \left\langle a \otimes b, \frac{\partial \nu P}{\partial x_i^{\varepsilon_j}} \otimes \frac{\partial \nu Q}{\partial x_q^{\varepsilon_l}} \{x_i^{\varepsilon_j}, x_q^{\varepsilon_l}\} \right\rangle.$$

Recall that the codimension of a generic orbit (i.e., orbit of maximal dimension) is called the *index* of the Lie algebra. The following assertion allows us to calculate the index of the extended algebra G_A^* via information about the original algebra G^* and the algebra A.

THEOREM 3.2. *Let the number r of polynomial invariants of the algebra G^* be equal to the index of G^*. Then*

$$\mathrm{ind}(G_A^*) = \dim A \cdot \mathrm{ind}(G^*).$$

Note. It is not possible to exhibit a collection of polynomial invariants equal in number to the index for an arbitrary Lie algebra. However, this is possible for algebraic Lie algebras.

The fundamental result of this section is the following:

THEOREM 3.3. *Let $P_1, \ldots, P_s \in k[x_1, \ldots, x_m]$ form a complete (i.e. maximal) commutative collection of functions (polynomials) on G^*. Then the polynomials $P_i^{\varepsilon_j}$ constructed above form a commutative collection of functions on the extended algebra G_A^*.*

In other words, the invariants of the algebra G_A^* have to be translated by elements $a = \varepsilon_j$.

Now let a Lie algebra G be given, and suppose that a splitting $G = H + V$ is given such that $[H, H] \subset H$, $[H, V] \subset V$, and $[V, V] \subset H$, i.e. the splitting corresponding to a symmetric space with group $\mathfrak{G}$. Such Lie algebras are sometimes called $\mathbf{Z}_2$-*graded*. The results which follow below are due to A. V. Brailov.

Let $G^* = H^* \oplus V^*$ be the dual splitting. If $x \in G^*$ and $g \in G$, then we denote the H^*, V^*-components of x and H, V-components of g by x_H, x_V, g_H, and g_V respectively. Define on the algebra the bilinear operation

$$[g, g']_\lambda = [g_H, g'_H] + [g_V, g'_H] + [g_H, g'_V] + \lambda^2 [g_V, g'_V]$$

for $g, g' \in G$. This operation satisfies the Jacobi identity, so we have obtained some new Lie algebra which we denote by G_λ. By definition we call the algebra G_0 a *contraction* of the $\mathbf{Z}_2$-graded Lie algebra G.

Along with contraction of an algebra we also define contractions of functions.

Let F be a smooth function on G^*. Assume that the decomposition of the function according to degrees of the V-variables terminates (for example, F is a polynomial):

$$F(x) = F^0(x_H, x_V) + F^1(x_H, x_V) + \cdots + F^n(x_H, x_V).$$

Let $F_\lambda(x) = \lambda^n F(x_H, \lambda^{-1} x_V)$. Since

$$\lambda^n F^n(x_H, \lambda^{-1} x_V) = F^n(x_H, x_V),$$

$F_\lambda(x)$ is defined also for $\lambda = 0$. We call the function $F_0(x)$ a *contraction* of F.

LEMMA. *Let F and F' be two functions on G^* which are in involution. If their contractions F_0 and F'_0 are defined (as will always be the case if the original functions are polynomials), then they are also in involution on the "contracted algebra" G_0^*, i.e. $[F_0, F'_0]_0 = 0$. (See above for the definition of the commutation operation.)*

PROPOSITION 3.1. *Let F be a polynomial on G^*, invariant with respect to the coadjoint representation (i.e. an invariant of the algebra); then its contraction F_0 is an invariant of the coadjoint representation of the Lie group corresponding to the Lie algebra G_0 (i.e. is an invariant of G_0).*

Let G be a real Lie algebra with $G^{\mathbf{C}}$ its complexification, let Σ be a group acting by automorphisms on $G^{\mathbf{C}}$, let G_n be a Σ-fixed subalgebra of G, and let $a \in G^*$. Suppose a is a characteristic covector of weight χ relative to the representation of the compact group Σ on the algebra $(G^{\mathbf{C}})^*$. Any symmetric operator $\varphi \colon G_n^* \to G_n$ for which there is an element $b \in G$ such that the identities

$$\{a, b\} = 0 \quad \text{and} \quad \{a, \varphi(x)\} = \{x, b\}, \qquad x \in G_n^*,$$

are satisfied, we shall call an (a, b)-*operator*. Now take as G the following algebra: $u(p, q) \times \mathbf{C}^{p+q}$, i.e. the semidirect product of the algebra $u(p, q)$ with the standard representation of minimal dimension. Then $G^{\mathbf{C}}$ is obtained as a contraction from the algebra $\mathrm{sl}(n, \mathbf{C})$.

THEOREM 3.4. *For any generic covector $a \in G^*$ the Euler equations $\dot{x} = \{x, \varphi(x)\}$ are completely integrable (in the sense of Liouville) for any (a, b)-operator (see the description above), such operators existing for any generic covector a.*

This result extends the result obtained in [5] for the case of a semidirect product of an orthogonal Lie algebra with its standard representation of minimal dimension. Operators of (a, b)-series are analogs of the operators introduced in constructing systems of the form $\dot{X}_C$ (i.e. analogs of section operators). An analogous result is true also for Lie algebras $\mathrm{so}(p, q) \times \mathbf{R}^{p+q}$. Application of the same method will probably allow the same result to be obtained without particular difficulty for the algebra $\mathrm{sp}(n) \times \mathbf{C}^{2n}$ also.

References

1. A. T. Fomenko, *Algebraic structure of certain classes of completely integrable Hamiltonian systems on Lie algebras*, Geometric Theory of Functions and Topology (N. P. Korneĭchuk, editor), Inst. Mat. Akad. Nauk Ukrain. SSR, Kiev, 1981, pp. 85–126. (Russian)

2. A. S. Mishchenko and A. T. Fomenko, *Euler equations on finite-dimensional Lie groups*, Izv. Akad. Nauk SSSR Ser. Mat. **42** (1978), 396–415; English transl. in Math. USSR Izv. **12** (1978).

3 ____, *A generalized Liouville method for the integration of Hamiltonian systems*, Funktsional. Anal. i Prilozhen. **12** (1978), no. 2, 46–56; English transl. in Functional Anal. Appl. **12** (1978).

4. A. T. Fomenko, *Group symplectic structures on homogeneous spaces*, Dokl. Akad. Nauk SSSR **253** (1980), 1062–1067; English transl. in Soviet Math. Dokl. **22** (1980).

5. V. V. Trofimov and A. T. Fomenko, *A method for constructing Hamiltonian flows on symmetric spaces and the integrability of some hydrodynamical systems*, Dokl. Akad. Nauk SSSR **254** (1980), 1349–1353; English transl. in Soviet Math. Dokl. **22** (1980).

6. S. M. Vishik and F. V. Dolzhanskiĭ, *Analogs of the Euler-Lagrange equations and magnetohydrodynamics connected with Lie groups*, Dokl. Akad. Nauk SSSR **238** (1978), 1032–1035; English transl. in Soviet Math. Dokl. **19** (1978).

7. A. A. Arkhangel'skiĭ, *Completely integrable Hamiltonian systems on a group of triangular matrices*, Mat. Sb. **108(150)** (1979), 134–142; English transl. in Math. USSR Sb. **36** (1980).

8. V. V. Trofimov, *Euler equations on Borel subalgebras of semisimple Lie algebras*, Izv. Akad. Nauk SSSR Ser. Mat. **43** (1979), 714–732; English transl. in Math. USSR Izv. **14** (1980).

9. ____, *Finite-dimensional representations of Lie algebras and completely integrable systems*, Mat. Sb. **111(153)** (1980), 610–621; English transl. in Math. USSR Sb. **39** (1981).

10. N. E. Kochin, I. A. Kibel', and N. V. Roze, *Theoretical hydromechanics*. I, 6th ed., Fizmatgiz, Moscow, 1963; English transl. of 5th ed., Interscience, 1964.

11. V. A. Steklov, *On the motion of a solid body in a liquid*, Dappe, Kharkov, 1893. (Russian)*

12. G. V. Gorr, L. V. Kudryashova, and L. A. Stepanova, *Classical problems in the theory of solid bodies. Their development and current state*, "Naukova Dumka", Kiev, 1978. (Russian)

13. G.-H. Halphen, *Sur le mouvement d'un solide dans un liquide*, J. Math. Pures Appl. (4) **4** (1888), 5–81 (especially 28–37).

14. ____, *Traité des fonctions elliptiques et de leurs applications*. Vol. II, Gauthier-Villars, Paris, 1888.

15. H. Weber, *Anwendung der Thetafunctionen zweier Veränderlicher auf die Theorie der Bewegung eines festen Körpers in einer Flüssigkeit*, Math. Ann. **14** (1879), 173–206.

16. V. I. Arnol'd, *Mathematical methods in classical mechanics*, "Nauka", Moscow, 1974; English transl., Springer-Verlag, 1978.

Translated by D. R. CHILLINGWORTH

Editor's note.* The same subject matter is covered, although from a somewhat different viewpoint, in his paper *Mémoire sur le mouvement d'un corps solide dans un liquide indéfini*, Ann. Fac. Sci. Univ. Toulouse (2) **4 (1902), 171–219.

Amer. Math. Soc. Transl.
(2) Vol. **133**, 1986

A Conjecture about Entropy

A. B. KATOK

1. In the Introduction to [1], M. Shub cites a definition of the topological entropy of a continuous map T of a compact metric space in terms of the asymptotics of the number of elements in (n, ε)-separated sets. We supplement this with some remarks which will be useful in what follows.

We set

$$d_n(x, y) = \max_{0 \leq i \leq n} d(T^i x, T^i y)$$

and let $r_n(T, \varepsilon)$ be the minimum number of elements in an ε-net in the space X with metric d_n. It is clear that

$$r_n(T, \varepsilon/2) \geq S_n(T, \varepsilon) \geq r_n(T, \varepsilon), \tag{1}$$

where $S_n(T, \varepsilon)$ is the maximal number of elements in an (n, ε)-separated subset of X.

In fact, the right-hand inequality follows from the fact that a maximal (n, ε)-separated set in X is an ε-net with respect to the metric d_n, and the left holds because any (n, ε)-separated set of balls with respect to the metric d_n centered at the points of this set are pairwise disjoint and any $\varepsilon/2$-net must meet such a ball. Inequality (1) implies the following equivalent definition of the topological entropy:

$$h(T) = \lim_{\varepsilon \to 0} \varlimsup_{n \to \infty} \frac{1}{n} \log r_n(T, \varepsilon). \tag{2}$$

We let $B_n(x, \varepsilon)$ denote the ball in X centered at x and having radius ε with respect to the metric d_n.

In [1], Shub stated the following conjecture.

1980 *Mathematics Subject Classification* (1985 *Revision*). Primary 54H20, 58F11; Secondary 58F15.

Translation of Smooth Dynamical Systems (D. V. Anosov, editor), "Mir", Moscow, 1977, pp. 181–203; MR **58** #18596.

©1986 American Mathematical Society
0065-9290/86 $1.00 + $.25 per page

THE ENTROPY CONJECTURE (see [1], §V). *For any C^1 map f of a compact manifold M to itself,*

$$h(f) \geq \log s(f_*). \tag{3}$$

Here $f_* \colon H_*(M, \mathbf{R}) \to H_*(M, \mathbf{R})$ denotes the linear map induced by f on the total homology of M,

$$H_*(M, \mathbf{R}) = \bigoplus_{i=0}^{\dim M} H_i(M, \mathbf{R}), \tag{4}$$

and $s(f_*)$ is the spectral radius of f_*. That is, $s(f_*) = \lim_{n \to \infty}(\| f_*^n \|)^{1/n}$, which coincides with the maximum of the moduli of the eigenvalues of f_*.

We allow ourselves a little abuse of language and say that a map or class of maps *satisfies the entropy conjecture* if we can establish that (3) holds for such maps.

We will investigate partial results, counterexamples, and some suggestive considerations related to the entropy conjecture and its possible generalizations. We feel that the conjecture has been a very fruitful problem and that attempts to prove it have been very helpful for the development of that direction in the theory of dynamical systems, connected in the first place with the work of Smale, Shub, and Sullivan, which tends to unite the methods of investigation of smooth maps used in differential topology and the theory of smooth dynamical systems (differentiable dynamics).

The available partial results can be divided into three groups: assertions weaker than the entropy conjecture which have been proved for arbitrary smooth or even continuous maps, a proof of the entropy conjecture for special classes of manifolds, and a proof of it for special classes of maps. We begin with some general remarks, then examine the available results in the above order, and conclude with a discussion of when the equality $h(f) = \log s(f_*)$ is attained.

2. Since the expansion (4) is clearly invariant under f_*, we have

$$s(f_*) = \max_{1 \leq i \leq \dim M} s(f_{*i}),$$

where f_{*i} denotes the restriction of f_* to the space $H_i(M, \mathbf{R})$.

Thus, (3) is equivalent to the system of inequalities

$$h(f) \geq \log s(f_{*i}), \qquad i = 1, \ldots, \dim M. \tag{5}$$

Throughout what follows, we will suppose that the manifold M has a fixed Riemannian metric, and we let $d(x, y)$, $x, y \in M$, denote the distance function on M induced by this metric. Bounding from above the action of f on homology (and this is necessary for the entropy conjecture) can be carried out by the following considerations.

We shall consider k-dimensional chains generated by smooth singular simplexes. If σ^k is such a chain, let $\lambda_k(\sigma^k)$ denote its k-dimensional Riemannian volume. Define the volume of the k-dimensional chain $c^k = \sum a_i \sigma_i^k$, $a_i \in \mathbf{R}$, to be

$\lambda_k(c^k) = \sum_i |a_i| \lambda_k(\sigma_i^k)$. Define the norm $\|\alpha\|$ of a homology class $\alpha \in H_k(M, \mathbf{R})$ to be the infimum of the volumes of all cycles representing the class α. It is clear that $\|\alpha + \beta\| \leq \|\alpha\| + \|\beta\|$ and that $\|a\alpha\| = |a| \cdot \|\alpha\|$ for $a \in \mathbf{R}$. In order to verify that $\| \ \|$ actually defines a norm on $H_k(M, \mathbf{R})$, we need only check that $\|\alpha\| = 0$ implies $\alpha = 0$.

To do this, observe that for any k-dimensional differential form γ, the inequality $\int_{\sigma^k} \gamma \leq c(\gamma) \lambda_k(\sigma^k)$ holds for some constant $c(\gamma)$. Choose a basis of $H^k(M, \mathbf{R})$ and represent the elements of the basis by differential forms $\gamma_1, \ldots, \gamma_s$. If $\|\alpha\| = 0$, then for any $\varepsilon > 0$ there exist cycles $c_\varepsilon \in \alpha$ such that $|\int_{c_\varepsilon} \gamma_i| < \varepsilon$ for $i = 1, \ldots, s$. But, since these integrals do not depend on the choice of cycles representing α, the de Rham theorem implies that $\alpha = 0$.

Thus, to establish the inequality $\log s(f_{*k}) \leq h(f)$ it suffices to show that for any sufficiently small, smooth, singular k-dimensional simplex σ^k there is a chain c_n homologous to $f^n \sigma^k$ such that

$$\lim_{n \to \infty} \frac{\log \lambda_k(c_n)}{n} \leq h(f).$$

In particular, this inequality holds if we have

$$\overline{\lim_{n \to \infty}} \frac{\log \lambda_k(f^n \sigma^k)}{n} \leq h(f) \tag{6}$$

for a generic simplex σ^k.

Of course, when f is not one-to-one the quantity $\lambda_k(f^n \sigma^k)$ must be computed counting multiplicity. The stipulation following (6) that the simplex be generic is crucial. As G. A. Margulis has remarked, without this stipulation, inequality (6) can fail even for Morse-Smale diffeomorphisms. It is true that in his examples either the diffeomorphisms or the simplexes are only finitely differentiable. We note that inequality (6) pertains exclusively to differentiable dynamics, and were one to successfully prove it, then the entropy conjecture would be established without recourse to the methods of differential topology.

3. THEOREM 1 (A. MANNING [2]). *If f is any continuous map of a smooth compact manifold M, then*

$$h(f) \geq \log s(f_{*1}).$$

PROOF. The smooth one-dimensional simplexes are just the paths on M. We let $*$ denote the usual composition of paths. A homotopy of a path which is not closed shall be understood to be a homotopy which fixes the endpoints. By the discussion above, it suffices to show that for any sufficiently small path σ and for some $\varepsilon > 0$, the image $f^n \sigma$ is homotopic to a path whose length is bounded by a constant multiple of $r_n(f, \varepsilon)$.

Choose $\delta > 0$ such that any ball of radius δ on M is contractible. Choose $\varepsilon > 0$ so that any two points x and y whose distance from one another does not exceed 4ε can be joined by a path of length less than δ/K lying in a ball of radius δ/K, where $K = \max_{x \in M} \|Df_x\|$.

Now suppose that x and y are the endpoints of a path σ which lies entirely in a ball of radius ε. Let Q_n be an ε-net with respect to the metric d_n on M which consists of $r_n(f, \varepsilon)$ elements. Choose points $x_0, x_1, \ldots, x_s \in Q_n$ and points $z_0 = x, z_1, \ldots, z_s = y$ on σ such that $z_i \in B_n(x_i, \varepsilon)$ and the segment $\{z_i, z_{i+1}\}$ of the path σ between the points z_i and z_{i+1} lies in the union of the balls $B_n(x_i, \varepsilon)$ and $B_n(x_{i-1}, \varepsilon)$, $i = 0, 1, \ldots, s$. Connect the points z_i to x_i by means of a path $\{z_i, x_i\}$ lying entirely in a ball of radius δ/K. Consider the path

$$\sigma' = \{z_0, x_0\} * \{x_0, z_0\} * \{z_0, z_1\} * \cdots * \{z_{s-1}, z_s\} * \{z_s, x_s\} * \{x_s, z_s\},$$

where $\{x_i, z_i\}$ denotes the path running in the opposite direction to $\{z_i, x_i\}$. It is clear that σ' is homotopic to σ. If two of the points $x_0, \ldots, x_s$ coincide then σ' contains a loop beginning and ending at these points; this loop is clearly null-homotopic. Eliminating such loops from σ' gives a new path σ'' which is homotopic to σ and σ' and consists of elements of the form $\{z_i, z_{i-1}\}$, $\{x_i, z_i\}$, and $\{z_i, x_i\}$. Moreover, the total number of such elements does not exceed $3r_n(f, \varepsilon)$. We show by induction that, for each member $\kappa = \{y, y'\}$ of the path σ'' and for each $i = 0, 1, \ldots, n$, the image $f^i \kappa$ is homotopic to a path κ_i of length less than δ/K which lies in a ball of radius δ/K. In fact, if $f^i \kappa$ is homotopic to κ_i, then $f^{i+1}\kappa$ is homotopic to $f\kappa_i$; this path lies in a contractible ball of radius δ and, consequently, is homotopic to a path κ_{i+1} of length less than δ/K which lies in a ball of radius δ/K. The latter exists because the distance between the points $f^{i+1}y$ and $f^{i+1}y'$ is less than 4ε. Thus, the path $f^n\sigma''$, and consequently the path $f^n\sigma$, is homotopic to a path of length no greater than $3\delta r_n(f, \varepsilon)$. This proves the theorem.

We remark that we have, in fact, proved a stronger assertion than Theorem 1.

With the same hypotheses as Theorem 1, suppose that $\gamma_1, \ldots, \gamma_m$ is a system of generators of the group $\pi_1(M)$, and let $\gamma \in \pi_1(M)$. Consider all possible representations of $f_* \gamma$ in the form

$$\gamma_1^{i_1} \gamma_2^{i_2} \cdots \gamma_m^{i_m} \gamma_1^{i_{m+1}} \cdots \gamma_m^{i_{km}}, \qquad i_j \in \mathbf{Z}$$

(where f_* is the endomorphism of $\pi_1(M)$ induced by f) and let $\varphi_n(\gamma)$ be the minimum of $\sum_1^{km} |i_j|$ over all such representations. Then

$$\varlimsup_{n \to \infty} \frac{\log \varphi_n(\gamma)}{n} \le h(f).$$

Another way to generalize Theorem 1 is to relax the restrictions on the space on which f acts.

It is shown in [2] that it is not necessary to suppose that M is a manifold. It suffices that the following conditions be met:

1. For any $\varepsilon > 0$, there exists a $\delta > 0$ such that any two points x and y for which $d(x, y) < \delta$ can be joined by a path of diameter less than ε.

2. There exists an $\varepsilon_0 > 0$ such that any loop of diameter less than ε_0 is contractible in M.

4. The following useful remark pertains to the fact that we may, without loss of generality, assume that the manifold M is orientable when dealing with considerations relating to the entropy conjecture.

PROPOSITION 1. *Let M be a nonorientable manifold and $f\colon M \to M$ a continuous map. Let $\tilde{M}$ be the two-sheeted orientable cover of M and $\tilde{f}\colon \tilde{M} \to \tilde{M}$ the map covering f. Then $h(f) = h(\tilde{f})$ and $s(f_{*i}) \leq s(\tilde{f}_{*i})$ for $i = 1,\ldots,\dim M$.*

PROOF. We let $\pi\colon \tilde{M} \to M$ denote the projection and assume that the Riemannian metric on $\tilde{M}$ is induced from that on M. Clearly, $h(f) \leq h(\tilde{f})$. On the other hand, if ε is sufficiently small and if Q_n is an ε-net for the map f with respect to the metric d_n, then $\pi^{-1}(Q_n)$ is obviously an ε-net for the map $\tilde{f}$ with respect to the metric d_n. Thus $r_n(\tilde{f}, \varepsilon) \leq 2r_n(f, \varepsilon)$, and hence $h(\tilde{f}) = h(f)$.

The map $\pi_k^*\colon H^k(M, \mathbf{R}) \to H^k(\tilde{M}, \mathbf{R})$ on cohomology induced by f is injective, and $\pi_k^* f_k^* = \tilde{f}_k^* \pi_k^*$, where $f_k^*\colon H^k(M, \mathbf{R}) \to H^k(M, \mathbf{R})$ is the map induced by f. Since $s(f_{*k}) = s(f_k^*)$, the proposition follows.

When M is a compact orientable m-dimensional manifold, Poincaré duality gives maps $D_i\colon H_i(M, \mathbf{R}) \to H^{m-i}(M, \mathbf{R})$ which are defined as follows. If $\alpha \in H_i(M, \mathbf{R})$ and $\beta \in H_{m-i}(M, \mathbf{R})$, then

$$(D_i\alpha)(\beta) = \langle \alpha, \beta \rangle,$$

where $\langle \alpha, \beta \rangle$ denotes the intersection index of the cycles. Since

$$\langle f_{*i}\alpha, f_{*m-i}\beta \rangle = \deg f \langle \alpha, \beta \rangle,$$

we obtain the relation

$$f_{m-i}^* = \deg f D_i f_{*i}^{-1} D_i^{-1},$$

when $f\colon M \to M$ is a homeomorphism. Consequently,

$$s(f_{*m-i}) = s(f_{m-i}^{-1*}) = s(f_{i*}^{-1}). \tag{7}$$

Theorem 1, Proposition 1, formula (7), and the equality $h(f) = h(f^{-1})$ now imply the following assertion (see [2]).

COROLLARY 1. 1. *If $f\colon M \to M$ is a homeomorphism and $\dim M = m$, then $\log s(f_{*m-1}) \leq h(f)$.*

2. *The entropy conjecture holds for any homeomorphism of manifolds with dimension less than or equal to three.*

We mention the following assertion related to homeomorphisms which was formulated in [3] as "Theorem 2" (the quotation marks are those of the authors) and for which a heuristic proof was sketched.

If $\dim M \geq 5$, the entropy conjecture holds for all homeomorphisms belonging to an open dense set of the space of all homeomorphisms of M with the C^0-topology.

In [3], the authors propose to imitate for homeomorphisms the "Markov approximation" procedure for diffeomorphisms described by Shub and Sullivan in [4], §1, and thereby construct a dense set of homeomorphisms for which the entropy conjecture holds. They then assume that one can establish that such homeomorphisms are semistable.([1]) Nitecki in [5] established this for diffeomorphisms with a hyperbolic structure. This would imply that any sufficiently small perturbation in the C^0-topology of such a homeomorphism could not decrease the topological entropy.

For arbitrary homeomorphisms, the entropy conjecture does not hold (see Theorem 3, below).

5. The next case after the one-dimensional case in which inequality (5) has been successfully proved is for the top-dimensional homology groups; that is, the case when the dimension is equal to that of the manifold.

If $\dim M = m$, then $s(f_{*m}) = |\deg f|$, and thus the inequality $\log s(f_{*m}) \leq h(f)$ is trivially satisfied for homeomorphisms. Almost as trivial is the case when $f \colon M \to M$ is a continuous covering map. For, in this case, $|\deg f|$ is equal to the number of preimages of an arbitrary point $x \in M$ and the total preimage $\{f^{-n}x\}$ of x is an (n, ε)-separated set for any sufficiently small ε.

However, the inequality $h(f) \geq \log |\deg f|$ may fail to hold for an arbitrary continuous map. The simplest example of this sort was constructed by Shub in [6] and is based on the same idea as the suspension of the homeomorphism of cell complexes (not manifolds) described in §V of [1]. We consider the sphere S^n $(n \geq 2)$ as the suspension of S^{n-1} and we choose a map $g \colon S^{n-1} \to S^{n-1}$ such that $|\deg g| > 1$. Compose the suspension of g with a map which moves each point along a "meridian" from the "north pole" N of the sphere S^n towards the "south pole" S. As a result we obtain a map $f \colon S^n \to S^n$ satisfying $\deg f = \deg g$, whose nonwandering set $\Omega(f)$ consists of the poles N and S. Thus,

$$h(f) = h(f \mid \Omega) = 0 < \log |\deg f|.$$

We shall consider this example in a little more detail. If the map g is chosen to be smooth and if the map which pushes down along the meridians is smooth along the meridians, then f is trivially smooth everywhere except at the unstable pole N and the stable pole S. Moreover, in a neighborhood of the stable pole, we can make the map f a strict contraction. By the same token, it is possible to make f infinitely differentiable (and, in some cases, even analytic) at S. The index of f and all its iterates at this point is equal to 1.

It turns out that the map f fails to be smooth in an essential way at N. In fact, f is not even a local homeomorphism at this point. Thus, if we were able to "smooth" f at N, the Jacobian of f would have to be equal to zero in spite of the fact that N is a repelling point.

Another interpretation, due to Shub, of the essential nonsmoothness of f at N is (somewhat freely translated) as follows. A direct computation shows that the index of f^n at N is equal to $(\deg g)^n$. On the other hand, Shub and Sullivan proved in [7] that the index at a fixed point of any iterate of a smooth map is

---

([1])A map f is called *semistable* if, for any map g sufficiently close to f in the C^0-topology, there exists a continuous map h for which $f \circ h = h \circ g$.

bounded. Thus, were f to be a limit in the C^0-topology of smooth maps, the point N would have to "absorb" an infinite set of periodic points in the limit. This would guarantee smooth maps with sufficiently large topological entropy.

It turns out that two of the properties of smooth maps, boundedness of the Jacobian (the expansion coefficient of the Riemannian volume) and being local homeomorphisms at points where the Jacobian is different from zero, are already sufficient to prove the inequality $h(f) \geq \log |\deg f|$ by imitating the elementary arguments for covers cited in the beginning of this section. This result is due to Misiurewicz and Przytycki, who proved it first for two-dimensional manifolds in [8] and later for the general case, in a considerably simpler fashion, in [9]. We reproduce the latter proof here.

THEOREM 2 (MISIUREWICZ AND PRZYTYCKI). *If $f\colon M \to M$ is any C^1-map of a smooth compact manifold M, then $h(f) \geq \log |\deg f|$.*

PROOF. Suppose that $0 < \alpha < 1$ and let L be the lowest upper bound of the absolute value of the Jacobian of f. Set $\varepsilon = L^{-1/(1-\alpha)}$ and let B be the compact set for which the absolute value of the Jacobian is no less than ε. Cover B by open sets on which f is a local diffeomorphism. Let δ be the Lebesgue number of this cover. Thus, if $x, y \in B$ and $d(x, y) \leq \delta$, then $f(x) \neq f(y)$.

Fix a natural number n and consider the set $A = A_n$ consisting of those points x, y for which no more than αn of the images $x, f(x), \ldots, f^{n-1}(x)$ belong to B. If $x \in A_n$, then the absolute value $|Jf^n(x)|$ satisfies the estimate

$$|Jf^n(x)| = \prod_{j=0}^{n-1} |Jf(f^j(x))| < \varepsilon^{(1-\alpha)n} L^{\alpha n} \leq (\varepsilon^{1-\alpha} L)^n = 1$$

and, consequently, the Riemannian volume of the set $f^n A$ is less than the volume of the whole manifold M.

Using this fact and Sard's theorem, we choose a regular value $x \in M \backslash f^n(A)$ of f^n. Now choose a sufficiently large (n, δ)-separated set from among the n preimages of x. We argue as follows. Each regular value y of the map f has no more than $N = |\deg f|$ preimages. If among these preimages there are N preimages from the set B, we choose them. Otherwise, choose a preimage which does not belong to B. Beginning at the point x and inductively applying this procedure, we obtain in turn some subset of the set $f^{-1}(\{x\})$, then some subset of $f^{-2}(\{x\})$, and so forth, until we obtain a subset Q_n of $f^{-n}(\{x\})$. By the choice of δ and the description of the procedure, it is clear that Q_n is an (n, δ)-separated set. We bound from below the number of elements of this set. Suppose $y \in Q_n$. Since $x \notin f^n(A)$ and $y \in f^{-n}(\{x\})$, we have $y \notin A$. In the construction of Q_n, we considered two sorts of transitions to preimages: taking N "good" preimages or a single "bad" preimage. Since $y \notin A$, there are no more than αn numbers k between 0 and $n - 1$ for which $f^k(y) \in B$. This means that in passing from x to y there are at least $m = [\alpha n] + 1$ "good" transitions. Since this holds for any $y \in Q_n$, there are at least $N^m \geq N^{\alpha n}$ elements, and consequently $S_n(f, \delta) \geq N^{\alpha n}$, from which it follows that $h(f) \geq \alpha \log N$. Since α can be chosen arbitrarily close to 1, we have $h(f) \geq \log N$.

COROLLARY 2. *The entropy conjecture holds for any smooth map of*
1) *spheres S^n, or*
2) *any manifold of dimension no more than two.*([2])

6. We now turn to the case of invertible maps. Recall that here the case of m-dimensional homology is trivial ($m = \dim M$) and inequality (5) has already been proved for arbitrary homeomorphisms in the 1- and $(m-1)$-dimensional cases (Theorem 1 and Corollary 1). The situation turns out to be much more complicated in the intermediate dimensions. We cite a negative result obtained as a result of discussions at the Warwick symposium on dynamical systems in 1974. An account is given in a report by Pugh [10] (the title of which lists the participants in the discussions).

THEOREM 3 [10]. *There exists a homeomorphism f of a compact, smooth, 8-dimensional manifold M for which $\Omega(f)$ is a finite set and $s(f_{*2}) > 1$.*

Thus, $0 = h(f) < \log s(f_{*2})$ and the entropy conjecture does not hold for the homeomorphism f. The construction of f is based on a further development of the idea of suspension using some results and methods of piecewise linear topology.

We first construct a suspension K of the two-dimensional torus. By composing the suspension of an Anosov diffeomorphism of the torus with a translation map on K which moves points from the north pole to the south, we obtain a homeomorphism $B \colon K \to K$ for which the set $\Omega(B)$ consists of two points and for which $s(f_{*2}) > 1$.

We then construct a piecewise linear inclusion $i \colon K \to \mathbf{R}^8$. It turns out that the homeomorphism induced by B on the image iK can be extended to a homeomorphism $\overline{B}$ of the Euclidean space $\mathbf{R}^8$. Let N be the star neighborhood of iK in the second barycentric subdivision of a triangulation of $\mathbf{R}^8$ which contains the polyhedron iK. By composing the homeomorphism $\overline{B} \mid N$ with a homeomorphism $h \colon \overline{B}N \to N$ which is the identity on iK (and which exists by a result of Hirsch [11]), we obtain an extension C of the homeomorphism B to the manifold with boundary N. Now consider the double M of the manifold N with the map C. This map has two invariant sets K_+ and K_-. The manifold M possesses a smooth structure. Furthermore, it is possible to perturb the map C, without changing it on K_+ or K_-, so that the points are moved from one pole of M to

---

([2])Assertion 2) actually only uses Theorem 2 in the case of spheres. In fact, by Proposition 1, it suffices to restrict attention to orientable surfaces. Theorem 1 then gives the result for the one-dimensional homology. For manifolds of genus greater than 1, the modulus of an iterate of any map does not exceed 1. Finally, for the torus, other considerations yield a more general result pertaining to arbitrary continuous maps (see Theorem 4).

the other, repelled from the set K_-, and attracted to K_+. Thus, the homeomorphism f of M constructed in this way has a finite number of nonwandering points. Consequently its topological entropy is zero.

It remains only to verify that, under the inclusion of the polyhedron K_+ into M, the two-dimensional cycles on K_+ do not become trivial. But this follows because $\dim M > 2 \dim K_+ + 1$ and K_+ is a retract of the set $M \backslash K_-$.

Shub's explanation in terms of indexes (see §5) is also applicable to this example.

7. On some manifolds, the entropy conjecture turns out to hold for arbitrary continuous maps.

THEOREM 4 (MISIUREWICZ AND PRZYTYCKI [12]). *The entropy conjecture holds for any continuous map f of the m-dimensional torus $\mathbf{T}^m$.*

PROOF. For definiteness, we will assume that the torus $\mathbf{T}^m = \mathbf{R}^m/\mathbf{Z}^m$ is endowed with the fixed standard Euclidean metric.

We first establish the inequality,

$$h(f) \geq \log |\deg f|, \tag{8}$$

for any map of the torus. For the torus, we have

$$\deg f = \det f_{1*}. \tag{9}$$

We let $\pi_m \colon \mathbf{R}^m \to \mathbf{T}^m$ denote the standard projection and consider the map $\tilde{f} \colon \mathbf{R}^m \to \mathbf{R}^m$ covering f. Suppose that $\Delta_n \subset \mathbf{R}^m$ is the standard fundamental domain in $\mathbf{R}^m$ (the Euclidean cube). From (9) and the fact that f is homotopic to an algebraic endomorphism of the torus induced by the linear operator f_{1*} on $\mathbf{R}^m$, it follows that the volume of the set $\tilde{f}^n \Delta_n$ is no smaller than $|\deg f|^n$. Thus, by a theorem of Minkowski, there exists a point $x \in \mathbf{T}^m$ such that the set $f^n \Delta_n \cap \pi_m^{-1}(x) = Q_n(x)$ contains no less than $|\deg f|^n$ points. For each point $y \in Q_n(x)$, choose a point $z(y) \in \tilde{f}^{-n}(y)$ and set $K_n = \bigcup_{y \in Q_n(x)} \{\pi_m z(y)\}$. We shall prove that, for sufficiently small $\varepsilon > 0$, K_n is (n, ε)-separated. Suppose that $x_1, x_2 \in K_n$ and the distance between x_1 and x_2 is sufficiently small. Consider points $z_1 \in \pi_m^{-1}(x_1)$ and $z_2 \in \pi_m^{-1}(x_2)$ such that $d(z_1, z_2) = d(x_1, x_2)$. Since $\tilde{f}^n z_1, \tilde{f}^n z_2 \in \pi_m^{-1}(x)$ are distinct, we have $d(f^n z_1, f^n z_2) \geq 1$. Thus, there exists a constant δ_0 which depends only on f, and not on n, for which the inequality $\delta_0 < d(\tilde{f}^k z_1, \tilde{f}^k z_2) < 1/2$ is satisfied for some k, $1 \leq k \leq n-1$. But, then

$$d(f^k x_1, f^k x_2) = d(\pi_m \tilde{f}^k z_1, \pi_m \tilde{f}^k z_2) = d(\tilde{f}^k z_1, \tilde{f}^k z_2) > \delta_0$$

and, consequently, K_n is an (n_1, δ_0)-separated set.

In order to establish the inequality $h(f) \geq \log s(f_{*k})$ for any k, we proceed in a similar fashion. We fix a standard basis γ for the group $H_k(\mathbf{T}^m, \mathbf{R})$ and realize each element α of this basis by the standard inclusion $i_\alpha \colon \mathbf{T}^k \to \mathbf{T}^m$ of the k-dimensional "coordinate" torus into $\mathbf{T}^m$. Suppose that $h_\alpha \colon \mathbf{T}^m \to \mathbf{T}^k$ is the standard projection, so that $h_\alpha \circ i_\alpha = \mathrm{id}_{\mathbf{T}^k}$. The matrix of f_{*k}^n with respect to the basis γ has elements of the form $c_{\alpha\beta}^n = \deg(h_\beta f^n i_\alpha)$ where $\alpha, \beta \in \gamma$.

Using Minkowski's theorem again, we find a point $x \in \mathbf{T}^k$ for which the set $Q_n(x) = \tilde{h}_\beta \tilde{f^n i}_\alpha \Delta_k \cap \pi_k^{-1}(x)$ contains no fewer than $c_{\alpha\beta}$ elements. Then, for each $y \in Q_n(x)$, we choose a point $z(y) \in (h_\beta \tilde{f^n i}_\alpha)^{-1} y$. An argument similar to the one above shows that the set $K_n = \bigcup_{y \in Q_n(x)} i_\alpha \pi_k(z(y))$ is (n, ε)-separated for sufficiently small ε.

Since

$$\log s(f_*) \leq \lim \frac{\log \sum_{\alpha, \beta \in \gamma} |c_{\alpha,\beta}^n|}{n},$$

the inequality $h(f) \geq \log(f_{*k})$ follows from (8).

We remark that for any algebraic endomorphism f_A of the torus generated by the matrix $A \in \mathrm{GL}(n, \mathbf{Z})$, we have

$$h(f_A) = \log s(f_{A*}) = \sum_{\substack{\lambda \in \mathrm{sp} A \\ |\lambda| > 1}} \log |\lambda|.$$

Since any Anosov automorphism of the torus is topologically conjugate to an algebraic automorphism (see [13] and [14]), we obtain the following assertion.

COROLLARY 3. *If f is an Anosov diffeomorphism of the torus $\mathbf{T}^m$, then $h(f) = \log s(f_*)$.*

We also mention the following property. If the map f_{*s} is invertible and has no eigenvalues on the unit circle, then, by Proposition 2.1 of Franks [13], there exists a continuous map $h \colon \mathbf{T}^m \to \mathbf{T}^m$ such that $hf = gh$, where g is the algebraic automorphism of the torus generated by f_{*1}. Thus, $h(f) \geq h(g) = \log s(f_{*1})$.

We remark that the entropy conjecture is established in [12] for homeomorphisms of orientable manifolds of the form $\mathbf{T}^m \times X$, where $\dim X = n$ and $H_i(X, \mathbf{R}) = 0$ for $0 < i \leq (m + n)/2$.

The arguments used in constructing (n, ε)-separated sets in the preimages of points which were successfully applied in the proofs of Theorems 2 and 4, or modifications of these arguments, might also turn out to be useful in some other cases. For example, they might be useful in proving the following.

CONJECTURE. *If M is a manifold whose universal covering space is homeomorphic to Euclidean space, then any continuous map $f \colon M \to M$ satisfies the entropy conjecture.*

8. Shub formulated the entropy conjecture in connection with the problem of defining the simplest diffeomorphisms in each isotopy class of diffeomorphisms (see [1] and [4]). From this standpoint, it is important to prove the entropy conjecture for "good" (for example, structurally stable) diffeomorphisms. In [4], Shub and Sullivan described an open and dense (in the C^0-topology) subset of the set of structurally stable diffeomorphisms for which the entropy conjecture holds. This subset consisted of the diffeomorphisms which were "Markov-fitted with respect to some handle decomposition of the manifold M." The structure of such a diffeomorphism accords well with the structure of the cell complex generated by the handle decomposition. As a result, $s(f_*)$ can be computed in terms of the algebraic intersection matrices and $h(f)$ can be bounded from below in terms of matrices consisting of the moduli of the elements of the intersection matrices. It does not follow from this situation that the entropy conjecture holds for the diffeomorphisms constructed by Shub and Sullivan. However, these diffeomorphisms satisfy the conditions of a theorem of Bowen [15] establishing the entropy conjecture for Axiom A, no-cycles diffeomorphisms which have nonwandering sets of dimension zero.

Later, Shub and Williams [16] obtained a more general result by eliminating the restriction $\dim \Omega(f) = 0$ and thus proved the entropy conjecture for all well-known (and possibly all) Ω-stable diffeomorphisms.

THEOREM 5 (SHUB AND WILLIAMS [16], announced in [6]). *The entropy conjecture holds for any Axiom A, no-cycles diffeomorphism.*

PROOF. We follow the outline of the proof in [16] with one essential difference. Let k be the dimension of the unstable subfoliation on a basic set Ω_i. Shub and Williams used Markov decompositions (to compute the topological entropy) and an extension of the system of stable manifolds to a neighborhood of the basic set to compute the volume of the part of the image of a k-dimensional simplex lying in Ω_i. Instead of computing the topological entropy from the intersection matrix associated to the Markov decomposition, we compute it directly from the asymptotics of the number of elements in (n, ε)-separating sets; and instead of extending the stable manifolds (which is always a delicate matter), we extend semi-invariant systems of cones around the stable and unstable subspaces (a procedure that presents no difficulties).

Recall that if $f: M \to M$ is an axiom A, no-cycles diffeomorphism, there exists a filtration $M_0 \subset M_1 \subset \cdots \subset M_m = M$ such that $f(M_i) \subset \mathrm{Int}(M_i)$ and $\bigcap_{n \in \mathbf{Z}} f^n \overline{(M_i \backslash M_{i-1})} = \Omega_i$ is a basic set of f for $i = 0, \ldots, m$. Let

$$f_*^{(i)}: H_*(M_i, M_{i-1}, \mathbf{R}) \to H_*(M_i, M_{i-1}, \mathbf{R}),$$

$$f_{*j}^{(i)}: H_j(M_i, M_{i-1}, \mathbf{R}) \to H_j(M_i, M_{i-1}, \mathbf{R}) \tag{10}$$

denote the maps induced by f. It is known that

$$h(f) = \sup_i h(f \mid \Omega_i).$$

On the other hand, from the exact homology sequences of the pairs (M_i, M_{i-1}), $i = 0, \ldots, m$, it is not difficult to show that

$$s(f_*) \leq \sup_i s(f_*^{(i)}).$$

Thus, to prove Theorem 5, it suffices to establish that

$$h(f \mid \Omega_i) \geq \log s(f_*^{(i)})$$

for each i.

Furthermore, instead of using the relative homology of the pair (M_i, M_{i-1}) to compute $s(f_*^{(i)})$, we can use the homology of the pair (X, A) where $X = f^N M_i$,

$A = X \cap f^{-N} M_{i-1}$, and N is an arbitrary positive integer. We will choose N such that the set $X \backslash A$ is contained in a sufficiently small neighborhood U of Ω_i which will be specified below. Observe that the set $X \backslash A$ is convex with respect to trajectories. This means that if $x \in X \backslash A$ and $f^n x \in X \backslash A$ for some $n > 0$, then $f^k x \in X \backslash A$ for $k = 0, 1, \ldots, n - 1, n$. This property follows at once from the properties of the filtration.

Suppose that we are given a cone K_x in the tangent space $T_x M$ for each x in a neighborhood U of Ω_i. We say that a submanifold $N \subset U$ is *compatible* with the system of cones K_x if $T_x N \subset K_x$ for every $x \in N$.

At each point $x \in \Omega_i$ we construct cones $K_x^s \supset E_x^s$ and $K_x^u \supset E_x^u$ which are "narrow" enough so that

$$Df^{-1} K_x^s \subset \operatorname{Int} K_{f^{-1}x}^s \quad \text{and} \quad Df K_x^u \subset \operatorname{Int} K_{fx}^u. \tag{11}$$

We do this in such a way that K_x^s and K_x^u depend continuously on x.

For example, we can choose a sufficiently small number $\gamma > 0$ and set

$$K_x^s = \{v \in T_x M : v = v_1 + v_2, \ v_1 \in E_x^s, \ v_2 \in E_x^u, \ \|v_2\| \leq \gamma \|v_1\|\},$$
$$K_x^u = \{v \in T_x M : v = v_1 + v_2, \ v_1 \in E_x^s, \ v_2 \in E_x^u, \ \|v_1\| \leq \gamma \|v_2\|\}.$$

If U is chosen sufficiently small, then we can extend the system of cones to a neighborhood U so that the formulas (11) are satisfied for $x \in U$ (provided $f^{-1}x$ or fx, as the case may be, lies in U) and so that there exist constants c_1 and $\varepsilon > \delta > 0$ such that any k-dimensional submanifold N consistent with the system of cones K_x^u possesses the following properties.

1. The volume of the sphere of radius ε in N does not exceed 1.

2. If $x, y \in N$ and the distance $d_N(x, y)$ between x and y in the intrinsic metric on N does not exceed δ, then $d_N(x, y) \leq c_1 d(x, y)$.

We now prove that $h(f | \Omega_i) \geq \log s(f_{*k}^{(i)})$. Using a relative variant of the arguments in §2, we easily see that it suffices to show that

$$\varlimsup_{n \to \infty} \frac{\log \lambda_k (f^n \sigma^k \cap (X \backslash A))}{n} \leq h(f \mid \Omega_i), \tag{12}$$

for any sufficiently small k-dimensional simplex $\sigma^k \subset X \backslash A$.

In addition, we can restrict ourselves to those simplexes for which the tangent space at each point x intersects the cone K_x^s only at the origin. Because $X \backslash A$ is convex with respect to trajectories in this case, there exists an $s > 0$ such that the manifold $f^s \sigma^k \cap (X \backslash A)$ is compatible with the cones K_x^u. Set $c_2 = \max_{x \in M} \|Df_x\|$ and cover $f^s \sigma^k \cap (X \backslash A)$ by a finite number of balls of radius $\delta / 2c_2$ in the intrinsic metric. Let N be any such ball. We estimate the volume of the manifold $f^n N \cap (X \backslash A)$. Choose a system of points on $f^n N \cap (X \backslash A)$ with the property that the distance between any two such points in the metric on $f^n N$ is larger than ε. Let S denote the nth preimage of this set.

We show that S is an $(n, \delta / c_1 c_2)$-separating set. Suppose that $x, y \in S$ and set $a_l = d_{f^l N}(f^l x, f^l y)$. Since $x, y \in N$, we have $a_0 \leq \delta / c_2$. On the other hand, $a_n \geq \varepsilon > \delta$ and, clearly, $a_{l+1} < c_2 a_l$. Therefore, there exists an l, $0 \geq l \geq n$, such that $\delta/c_2 \leq a_l \leq \delta$. We may assume that δ and the neighborhood of $X \backslash A$ have been chosen so small that the δ-neighborhood of $X \backslash A$ is contained in a neighborhood U for which properties 1 and 2 hold for the system of cones K_x^u. In particular, we can use the compatibility of $f^l N \cap U$ with the system K_x^u and property 2 to conclude that

$$d(f^l x, f^l y) \geq \delta/c_1 c_2.$$

Now suppose that the set S contains a maximal number of points. Then $f^n S$ is an ε-net on the set $f^n N \cap (X \backslash A)$ in the metric on $f^n N$. Since the system of cones K_x^u possesses property 1, the number of elements in such an ε-net is no less than the k-dimensional volume of the set $f^n N \cap (X \backslash A)$. Thus,

$$\lambda_k(f^n N \cap (X \backslash A)) \leq r_n(f, \delta/c_1 c_2).$$

Inequality (12) follows.

We will not deal in as much detail with the proof of the inequality $h(f \mid \Omega_i) \geq \log s(f_{*j}^{(i)})$ for $j \neq k$.

In this case, it is shown in [16] that the inequality is always strict. When $j < k$, it is possible to show that the volume of a j-dimensional simplex cannot grow faster than the volume of a k-dimensional simplex since there are more "free" directions in which to expand. For $j > k$, it is also necessary to use the result just established when $j \leq n - k$ for the dual filtration

$$\overline{M \backslash M_{m-1}} \subset \overline{M \backslash M_{m-2}} \subset \cdots \subset \overline{M \backslash M_0} \subset M$$

of the diffeomorphism f^{-1}.

The required result follows from duality between the $(n - j)$-dimensional cohomology of the pair $(\overline{M \backslash M_{i-1}}, \overline{M \backslash M_i})$ and the j-dimensional homology of the pair (M_i, M_{i-1}) (see [15]).

9. In this section we consider the question of which diffeomorphisms satisfy the equation $h(f) = \log s(f_*)$ or its local variants. In the case of diffeomorphisms with a hyperbolic structure, it is possible to find natural sufficient conditions. Although these conditions are not necessary, they cannot "essentially" be dispensed with.

We begin with Anosov diffeomorphisms.

PROPOSITION 2. *If $f \colon M \to M$ is an Anosov diffeomorphism and if the unstable subfoliation E^u of the tangent bundle TM is orientable, then $h(f) = \log s(f_*)$.*

PROOF. By Theorem 5, it suffices to prove that $h(f) \leq \log s(f_*)$. For hyperbolic sets (and, in particular, for Anosov diffeomorphisms), one can compute the topological entropy from the asymptotics of the number $N_n(f)$ of periodic points of f of period n. More explicitly,

$$h(f) = \overline{\lim} \, \frac{\log N_n(f)}{n}.$$

Let $P_n = \{x \in M, f^n x = x\}$ and let $i_{f^n}(x)$ be the index of x as a fixed point of f^n.

By the Lefschetz formula, we have

$$L(f^n) = \sum_{x \in P_n} i_{f^n}(x) = \sum_{i=0}^{\dim M} (-1)^i \operatorname{tr}(f^n_{*i}).$$

Since all periodic points of f^n are hyperbolic, the index equals $+1$ or -1 and depends only on the dimension. Thus, the map Df^n_x either preserves or reverses the orientation on the invariant expanding subspace E^u_x. Since M is connected and the subfoliation E^u is orientable, the spaces E^u_x can be oriented in a consistent manner such that the differential $Df^n_x|E^u_x \colon E^u_x \to E^u_{f^n x}$ is either orientation-preserving at all points or orientation-reversing at all points. In particular, this applies to the fixed points of f^n and, thus, the indexes $i_{f^n}(x)$ are equal for every $x \in P_n$. That is, $|L(f^n)| = N_n(f)$. Furthermore,

$$|L(f^n)| = \left| \sum_{i=0}^{\dim M} (-1)^i \operatorname{tr}(f^n_{*i}) \right| \leq \sum_{i=0}^{\dim M} |\operatorname{tr}(f^n_{*i})| \leq \dim H_*(M, \mathbf{R})(s(f_*))^n.$$

Therefore,

$$\overline{\lim} \, \frac{\log N_n(f)}{n} \leq s(f_*).$$

This proves the proposition.

The question of whether the unstable foliation E^u of an Anosov diffeomorphism is always orientable has been studied for more than ten years and has not yet been solved. In [19], Smale referred to this question in connection with the problem of the rationality of the zeta function of an Anosov diffeomorphism. This latter problem was subsequently solved using Markov decompositions [22] in the more general setting of hyperbolic sets where orientability may fail to hold (see below).

CONJECTURE. *If $f \colon M \to M$ is an Anosov diffeomorphism, then $h(f) = \log s(f_*)$.*

We now suppose that both the invariant subfoliations E^u and E^s of a given Anosov diffeomorphism are orientable (this is equivalent to the orientability of E^u and the manifold M itself). In this case, it is possible to amplify Proposition 2. In fact, let k be the dimension of E^u. Then, we can find a functional $\bar{\alpha}$ on the k-dimensional differential forms on M which geometrically realizes a nonzero element $\alpha \in H_k(M, \mathbf{R})$ for which $f_{*k}\alpha = \lambda\alpha$ and $\log|\lambda| = h(f)$. The construction is a particular case of a construction due to Ruelle and Sullivan [17].

On the global stable submanifolds W^s of M, there exists a family of σ-finite Borel measures μ_{W^s} (in general, singular) possessing the following properties (see [23] and [18]).

1. The measure of any compact subset of W^s is finite.

2. The measures pass into one another under translation along the local unstable submanifolds.

3. $f\mu_{W^s} = \lambda^{-1}\mu_{fW^s}$, where $\log|\lambda| = h(f)$.

Now let ω^k be a k-dimensional differential form on M. We let $\bar{\alpha}$ denote the functional whose value on ω^k is calculated as follows. Cover M by small open sets which have a local product structure (see [13]) and use a partition of unity to represent ω^k as a sum of forms with support in these sets. In each such open set, integrate the form along each local unstable manifold (taking into account the orientation, which is assumed to be chosen consistently), and then integrate the resulting integrals with respect to the measure μ_{W^s} along any transverse local stable manifold. Add the resulting contributions from each local form to obtain the value $\bar{\alpha}(\omega^k)$. We shall show that $\bar{\alpha}$ is a cycle; that is, $\bar{\alpha}(\partial\omega^{k-1}) = 0$ for any $(k-1)$-form ω^{k-1}. Let $\omega^{k-1} = \sum_i \omega_i^{k-1}$, where the support of each ω_i^{k-1} lies in the interior of some open set U_i with a local product structure. It is clear that

$$\bar{\alpha}(\partial\omega^{k-1}) = \sum_i \bar{\alpha}(\partial\omega_i^{k-1}).$$

But $\bar{\alpha}(\partial\omega_i^{k-1})$ is equal to the integral of the values of the form ω_i^{k-1} on the boundaries of the local unstable manifolds in U_i. Since these boundaries lie outside the support of ω_i^{k-1}, we have $\bar{\alpha}(\partial\omega_i^{k-1}) = 0$.

Let α denote the homology class of $\bar{\alpha}$. We need to show that $f_{*k}\alpha = \lambda\alpha$, $\log|\lambda| = h(f)$, and $\alpha \neq 0$.

Suppose, for definiteness, that the diffeomorphism f preserves the orientations of E^u and E^s (the other case can be handled by making the obvious modifications to the argument following).

We compute the intersection index of $\bar{\alpha}$ with $(n-k)$-dimensional chains. To do this, it suffices to find the intersection index of $\bar{\alpha}$ with any sufficiently small singular simplex σ^{n-k} transverse to the subfoliation E^u. Suppose U is an open set which has a local product structure and which contains σ^{n-k}. The simplex σ^{n-k} intersects each local unstable manifold at no more than one point. Thus, it follows from the definition of $\bar{\alpha}$ that the absolute value of the intersection index $\langle\alpha, \sigma^{n-k}\rangle$ is equal to the measure μ_{W^s} of the projection (along the local unstable manifolds) of σ^{n-k} onto any local stable manifold in U. The intersection index takes a plus or minus sign according to whether the orientation of the projection does or does not coincide with the orientation of the local stable manifolds.

From property 3 of the measures μ_{W^s} it follows that

$$\langle\bar{\alpha}, f\sigma^{n-k}\rangle = \lambda^{-1}\langle\bar{\alpha}, \sigma^{n-k}\rangle.$$

Linearity of the intersection index implies that a similar equation holds for any smooth singular $(n-k)$-dimensional chain. Thus, for any $\gamma \in H_{n-k}(M, \mathbf{R})$, we have

$$\langle\alpha, f_{*n-k}\gamma\rangle = \lambda^{-1}\langle\alpha, \gamma\rangle.$$

Since $\langle\alpha, \gamma\rangle$ is equal to the value on γ of $D_k\alpha \in H^{n-k}(M, \mathbf{R})$, we have $f_{n-k}^* D_k\alpha = \lambda D_k\alpha$. Since $f_{*k} = D_k^{-1}f_{n-k}^* D_k$, it follows that $f_{*k}\alpha = \lambda\alpha$, where $\log|\lambda| = h(f)$. Interchanging the roles of the stable and unstable manifolds, we can construct an $(n-k)$-dimensional cycle $\bar{\beta}$. From the definitions it is easy to see that $\langle \alpha, \beta \rangle \neq 0$, where β denotes the cohomology class of $\bar{\beta}$. Thus, $\alpha \neq 0$ and $\beta \neq 0$.

We now pass to the more general class of Axiom A, no-cycles diffeomorphisms. In this case, there are local variants of the propositions cited above. The orientation condition for the foliation E^u on a basic set Ω_i is as follows: There exists an orientation of the subfoliation E^u on Ω_i such that for every $x \in \Omega_i$ the differential

$$Df_x \mid E_x^u : E_x^u \to E_{f_x}^u$$

is either simultaneously orientation-preserving or simultaneously orientation-reversing. In this case,

$$h(f \mid \Omega_i) = \log s(f_*^{(i)}).$$

Shub and Williams proved this in [16] by an index argument which carries over word for word to this case by using a relative variant of the Lefschetz formula. In the case when the stable subfoliation E^s on Ω_i also satisfies the analogue of the orientability condition, Ruelle and Sullivan [17] constructed an eigenvector in $H_k(X, A, \mathbf{R})$ with eigenvalue λ, where $\log |\lambda| = h(f \mid \Omega_i)$. Their construction is similar to the one above for Anosov diffeomorphisms. The measures μ_{W^s} on the stable manifolds are constructed using conditional measures induced by the invariant measure with maximal entropy on Ω_i.

We give an example to show that the orientability conditions are essential. Smale, in the now classical horseshoe example (see [19], §I.5, especially Figures 7 and 13), constructed a diffeomorphism of the two-dimensional sphere S^2 with completely disconnected basic set. In this case, the unstable foliation on the basic set naturally possesses infinitely many orientations, but no orientation compatible with the action of the diffeomorphism exists. The topological entropy in Smale's example is equal to $\log 2$, while the spectral radius of the induced operator on the homology (in both the absolute and relative cases) is equal to 1.

Plikin [20] subsequently constructed an Axiom A diffeomorphism of S^2 with a one-dimensional attracting basic set on which the unstable subfoliation is not orientable. In this example the topological entropy is also positive, while a neighborhood of the basic set is contractible and, therefore, the spectral radius of the operator on homology is also equal to 1.

Finally, we remark that even if the orientability condition is satisfied on all the basic sets, it can happen that $h(f) > \log s(f_*)$, because

$$s(f_*) < \max_i s(f_*^{(i)}).$$

Gibbons [24] gives an example of this type on the three-dimensional sphere in which there are two basic sets, an attracting and a repelling solenoid, and the topological entropy is positive. The nontrivial relative one-dimensional cycles vanish under passage to absolute homology.

10. We conclude by mentioning some other unsolved problems connected with the topological entropy. In [4] it was shown that some isotopy classes of diffeomorphisms may fail to contain an Axiom A diffeomorphism f satisfying the strong transversality condition and the equality $h(f) = \log s(f_*)$.

PROBLEM. *Does every isotopy class of diffeomorphisms contain a diffeomorphism f for which $h(f) = \log s(f_*)$? In particular, does each isotopy class of diffeomorphisms for which $s(f_*) = 1$ contain a diffeomorphism with topological entropy equal to zero?*

It is known [25] that the topological entropy is neither continuous nor even upper or lower semicontinuous on $\mathrm{Diff}^r(M)$.

PROBLEM (see [**21**], Problem 41). *Is the topological entropy continuous on a second category set in $\mathrm{Diff}^r(M)$?*

We have already mentioned that the topological entropy of "good" diffeomorphisms can be computed in terms of the asymptotics of the number of periodic points. This is not true for arbitrary diffeomorphisms (see [1]).

PROBLEM. *Does the inequality*

$$h(f) \leq \varlimsup_{n \to \infty} \frac{\log N_n(f)}{n}$$

hold for diffeomorphisms in a second category set in $\mathrm{Diff}^r(M)$?

Many interesting unsolved problems pertaining to smooth dynamical systems are contained in the list of 50 problems compiled by Palis and Pugh [21]. These problems reflect the main directions in the theory of dynamical systems and the various questions discussed at the symposium at the University of Warwick in 1974.

REFERENCES

1. M. Shub, *Dynamical systems, filtrations and entropy*, Bull. Amer. Math. Soc. **80** (1974), 27–41.

2. Anthony Manning, *Topological entropy and the first homology group*, in [26], pp. 185–190.

3. J. Palis et al., *Genericity theorems in topological dynamics*, in [26], pp. 241–250.

4. M. Shub and D. Sullivan, *Homology theory and dynamical systems*, Topology **14** (1975), 109–132.

5. Zbigniew Nitecki, *On semi-stability for diffeomorphisms*, Invent. Math. **14** (1971), 83–122.

6. M. Shub, *Topological entropy and stability*, in [26], pp. 39–40.

7. M. Shub and D. Sullivan, *A remark on the Lefschetz fixed point formula for differentiable maps*, Topology **13** (1974), 189–191.

8. Michal Misiurewicz and Feliks Przytycki, *Entropy and degree for transformations of two-dimensional manifolds*, preprint (22 pp.).

9 ____ , *Topological entropy and degree of smooth mappings*, Bull. Acad. Polon. Sci. Sér. Sci. Math. Astr. Phys. **25** (1977), 573–574.

10. Charles Pugh, *On the entropy conjecture: a report on conversations among R. Bowen, M. Hirsch, A. Manning, C. Pugh, B. Sanderson, M. Shub, and R. Williams*, in [26], pp. 257–261.

11. Morris W. Hirsch, *Smooth regular neighborhoods*, Ann. of Math. (2) **76** (1962), 524–530.

12. Michal Misiurewicz and Feliks Przytycki, *Entropy conjecture for tori*, Bull. Acad. Polon. Sci. Sér. Sci. Math. Astr. Phys. **25** (1977), 575–578.

13. John Franks, *Anosov diffeomorphisms*, Global Analysis, Proc. Sympos. Pure Math., vol. 14, Amer. Math. Soc., Providence, R.I., 1970, pp. 61–93.

14. Anthony Manning, *There are no new Anosov diffeomorphisms on tori*, Amer. J. Math. **96** (1974), 422–429.

15. Rufus Bowen, *Entropy versus homology for certain diffeomorphisms*, Topology **13** (1974), 61–67.

16. Mike Shub and Robert F. Williams, *Entropy and stability*, Topology **14** (1975), 329–338.

17. D. Sullivan, *Homology classes composed of infinitely many unstable manifolds of a dynamical system*, in [26], pp. 42–44.

18. G. A. Margulis, *Some measures connected with Anosov flows on compact manifolds*, Funktsional. Anal. i Prilozhen **4** (1970), no. 1, 62–76; English transl. in Functional Anal. Appl. **4** (1970).

19. S. Smale, *Differentiable dynamical systems*, Bull. Amer. Math. Soc. **73** (1967), 747–817.

20. R. V. Plykin, *Sources and sinks of A-diffeomorphisms of surfaces*, Mat. Sb. **94(136)** (1974), 243–264; English transl. in Math. USSR Sb. **23** (1974).

21. J. Palis and C. C. Pugh (compilers), *Fifty problems in dynamical systems*, in [26], pp. 345–353.

22. Anthony Manning, *Axiom A diffeomorphisms have rational zeta functions*, Bull. London Math. Soc. **3** (1971), 215–220.

23. Ya. G. Sinaĭ, *Markov partitions and Anosov diffeomorphisms*, Funktsional. Anal. i Prilozhen. **2** (1968), no. 1, 64–89; English transl. in Functional Anal. Appl. **2** (1968).

24. Joel C. Gibbons, *One-dimensional basic sets in the three-sphere*, Trans. Amer. Math. Soc. **164** (1972), 163–178.

25. M. Misiurewicz, *Diffeomorphism without any measure with maximal entropy*, Bull. Acad. Polon. Sci. Sér. Sci. Math. Astr. Phys. **21** (1973), 903–910.

26. Anthony Manning (editor), *Dynamical systems—Warwick 1974* (Proc. Sympos., Coventry, 1973/1974, Presented to E. C. Zeeman on His Fiftieth Birthday), Lecture Notes in Math., vol. 468, Springer-Verlag, 1975.

Translated by D. B. O'SHEA

Amer. Math. Soc. Transl.
(2) Vol. **133**, 1986

Approximation by Polynomials of Solutions of a Class of Hammerstein Integral Equations

UDC 517.946

V. I. BILENKO

Abstract. An algorithm is determined for constructing polynomial solutions of Hammerstein integral equations with kernels of the form of the Green's function of a boundary value problem for ordinary differential equations. Conditions are established for the existence and uniqueness of such a solution. Efficient error estimates are obtained. A concrete computational algorithm is investigated in detail, and its efficiency in comparison with other algorithms is illustrated by using model examples.
Bibliography: 20 titles.

Introduction

In this paper we investigate an algorithm for constructing polynomial solutions of the Hammerstein integral equations

$$y(x) = f(x) + \int_a^b K(x,t)F(t,y(t))\,dt, \qquad x \in [a,b], \tag{1}$$

where $y(t)$ is the unknown solution, f and F are given functions, and the kernel has the form of the Green's function of a boundary value problem for ordinary differential equations, i.e.,

$$K(x,t) = \begin{cases} K_1(x,t) & \text{for } a \le t \le x \le b, \\ K_2(x,t) & \text{for } a \le x \le t \le b. \end{cases} \tag{2}$$

The equations (1) with such kernels are of interest because, first, the indicated boundary value problems reduce to them (see [1] and [2]), and, second, they are models for a class of processes essentially broader than boundary value problems. For example, the problem on radiation in a cloud cover, which does not have an

1980 *Mathematics Subject Classification* (1985 *Revision*). Primary 41A10, 45G10; Secondary 41A05, 41A25, 45-04, 65R20, 65V05.

Translation of preprint no. 80.17, Inst. Mat. Akad. Nauk Ukrain. SSR, Kiev, 1980; MR **82e:**65131.

©1986 American Mathematical Society
0065-9290/86 $1.00 + $.25 per page

equivalent in the form of a boundary value problem, reduces to such equations [3].

It is known (see [1]–[4]) that the existing algorithms for approximate solution of equations (1) and (2) become unwieldy, as a rule, when high accuracy is required. The algorithm proposed in this paper is to a significant degree free of this deficiency. The algorithm, and also its theoretical justification, are obtained by an approximation method for solving differential and integration equations which was proposed and developed by Dzyadyk in [5] and [6] and successfully used for solving various problems in articles of Denisenko [7], [8], Podlipenko [9], Gabdulkhaev [10], and others.

The algorithm was realized as a computer program. Its efficiency was illustrated by comparing the results of solving model examples by this algorithm with results obtained by other methods.

§1. Construction of approximating polynomials. An existence theorem

We assume that the functions f, K, and F have the following properties:

1) $f(x) \in C_{[a,b]}$, and $F(x, z)$ is continuous and satisfies a Lipschitz condition in the second variable with a constant A, i.e.,

$$|F(x, z') - F(x, z'')| \le A|z' - z''| \quad \text{for } (x, z') \text{ and } (x, z'') \in [a, b] \times [-Q, Q],$$

where Q is a sufficiently large constant.

2) The kernel $K(x, t)$ has the property that its averaged modulus of continuity $\omega_{CL}(K, \delta)$, which is defined $\forall \delta \in [0, b - a]$ by the formula (see [5])

$$\omega_{CL}(K, \delta) = (b - a)^{-1} \max_{\substack{0 \le h \le \delta \\ a \le x \le b-h}} \int_a^b |K(x + h, t) - K(x, t)|\, dt, \tag{3}$$

tends to zero as $\delta \to 0$: $\lim_{\delta \to 0} \omega_{CL}(K, \delta) = 0$.

3) Equation (1) has a unique solution.

We remark that property 2) is satisfied [5], for example, for kernels that generate compact integral operators acting from L_p ($p \ge 1$) (or $C_{[a,b]}$) to $C_{[a,b]}$.

Sufficient conditions for the existence and uniqueness of a solution of (1) are well known in the literature (see, for example, [1], [2], [11], and other references).

To construct an efficient algorithm for obtaining polynomials which are good approximations of the solution of (1) in the uniform metric (an algorithm that is suitable for realization on a computer), we consider the summator polynomial operators $\mathcal{U}_n$ (see [6], [7], and [12]) defined on $C_{[a,b]}$ by

$$\mathcal{U}_n(\psi; x) = \sum_{i=1}^N \psi(x_i)\pi_i(x) \stackrel{\mathrm{df}}{=} \mathcal{U}_n(\{\psi(x_i)\}_{i=1}^N; x) \quad \forall \psi \in C_{[a,b]}, \tag{4}$$

where $\pi_i(x) \stackrel{\mathrm{df}}{=} \sum_{k=0}^n a_{ik}\varphi_k(x)$ are standard generalized polynomials in some system of linearly independent functions $\{\varphi_k(x)\}_0^n \in C_{[a,b]}$.

It is assumed that the equality (3.21) in [5] holds for these operators:

$$\lim_{n\to\infty} \varepsilon(\mathcal{U}_n, H_K^\omega) = 0, \tag{5}$$

where

$$\varepsilon(\mathcal{U}_n, H_K^\omega) = \sup_{\psi \in H_K^\omega} \|\psi(x) - \mathcal{U}_n(\psi; x)\|_{C_{[a,b]}},$$

$$MH_K^\omega \overset{\mathrm{df}}{=} \{\psi \colon |\psi(x') - \psi(x'')| \le M\omega_{CL}(K,\delta)\} \tag{6}$$

$\forall x', x'' \in [a, b]$ such that $|x'' - x'| \le \delta \ \forall \delta \in [0, b - a]$, and $1H_K^\omega \overset{\mathrm{df}}{=} H_K^\omega$.

The most important examples of summator operators are the operators assigning to each continuous function the corresponding Lagrange interpolation polynomials, Bernstein polynomials, interpolation splines, or discrete analogues of various summation methods for Fourier-Tchebycheff series (Fejér polynomials, Vallée-Poussin polynomials, etc.).

The scheme for constructing polynomial solutions of (1) and (2) consists in the following.

1) Equation (1) is replaced according to [5] and [6] by the operator equation

$$\overset{\circ}{y}_n(x) = \mathcal{U}_n(f; x) + \mathcal{U}_n\left(\left\{\int_a^{x_i} K_1(x_i, t) F(t, \overset{\circ}{y}_n(t))\, dt \right.\right.$$
$$\left.\left. + \int_{x_i}^b K_2(x_i, t) F(t, \overset{\circ}{y}_n(t))\, dt \right\}_{i=1}^N ; x \right) \tag{7}$$

where $\overset{\circ}{y}_n(x)$ is a solution of this equation and $\mathcal{U}_n$ is an arbitrary summator operator of the form (4).

2) Each integral in (7) is computed by the quadrature formulas

$$\int_a^{x_i} \Phi_1(x_i, t)\, dt = \frac{x_i - a}{2} \sum_{j=1}^{m_1} p_j^{(1)} \Phi_1(x_i, t_{ij}^{(1)}) + R_{m_1 i}[\Phi_1] \overset{\mathrm{df}}{=} S_{m_1 i}[\Phi_1] + R_{m_1 i}[\Phi_1],$$

$$\int_{x_i}^b \Phi_2(x_i, t)\, dt = \frac{b - x_i}{2} \sum_{j=1}^{m_2} p_j^{(2)} \Phi_2(x_i, t_{ij}^{(2)}) + R_{m_2 i}[\Phi_1] \overset{\mathrm{df}}{=} S_{m_2 i}[\Phi_2] + R_{m_2 i}[\Phi_2],$$

$$\tag{8}$$

where $\{p_j^{(1)}\}_1^{m_1}$ and $\{p_j^{(2)}\}_1^{m_2}$ are the weights of the "standard" quadrature formulas S_{m_1} and S_{m_2} for the interval $[-1, 1]$, while $t_{ij}^{(1)}$ and $t_{ij}^{(2)}$ are the nodes of the quadrature formulas similar to them [5] on the intervals $[a, x_i]$ and $[x_i, b]$, and $R_{m_1 i}[\cdot]$ and $R_{m_2 i}[\cdot]$ denote the remainders of these formulas.

This reduces equation (7) to the "perturbed" (see [11], §17.1) equation

$$y_n(x) = \mathcal{U}_n(f; x) + \mathcal{U}_n\left(\{S_{m_1 i}[K_1 F(y_n)] + S_{m_2 i}[K_2 F(y_n)]\}_{i=1}^N ; x\right), \tag{7^1}$$

where

$$K_\nu F(z) \overset{\mathrm{df}}{=} K_\nu(x_i, t) F(t, z(t)), \qquad \nu = 1, 2. \tag{9}$$

3) On the basis of (4) and (8) the coefficients $\{c_k\}_0^n$ of a polynomial

$$y_n(x) = \sum_{k=0}^{n} c_k \varphi_k(x) \tag{10}$$

which is a solution of (7^1) are found by the method of successive approximations from the following system of transcendental equations:

$$c_k = \sum_{i=1}^{N} a_{ik} \left\{ f(x_i) + \frac{x_i - a}{2} \sum_{j=1}^{m_1} p_j^{(1)} K_1(x_i, t_{ij}^{(1)}) F\left(t_{ij}^{(1)}, \sum_{\nu=0}^{n} c_\nu \varphi_\nu(t_{ij}^{(1)})\right) \right.$$

$$\left. + \frac{b - x_i}{2} \sum_{j=1}^{m_2} p_j^{(2)} K_2(x_i, t_{ij}^{(2)}) F\left(t_{ij}^{(2)}; \sum_{\nu=0}^{n} c_\nu \varphi_\nu(t_{ij}^{(2)})\right) \right\}. \tag{11}$$

THEOREM 1. *Suppose that the summator operator $\mathcal{U}_n$ of the form (4), the quadrature formulas (8), and the functions F and K are such that*

$$A[\varepsilon(\mathcal{U}_n, H^{\omega_{11}})g_1 + \varepsilon(\mathcal{U}_n, H^{\omega_{12}})g_2 + (b - a)(g_1\|K_1\| + g_2\|K_2\|)] \stackrel{\mathrm{df}}{=} q \le 1, \tag{12}$$

where for $\nu = 1$ or 2 we write

$$\omega_{1\nu} = \omega_{1\nu}(K_\nu, \delta) \stackrel{\mathrm{df}}{=} [(b - a)\omega_1(K_\nu, \delta) + \delta\|K_\nu\|] \quad \forall \delta \in [0, b - a], \tag{13}$$

$$\omega_1(K_\nu, \delta) \stackrel{\mathrm{df}}{=} \sup_{a \le t \le b} \sup_{\substack{a \le x, x+h \le b \\ |h| \le \delta}} |K_\nu(x + h, t) - K_\nu(x, t)|, \tag{14}$$

$$MH^{\omega_{1\nu}} \stackrel{\mathrm{df}}{=} \{\psi \colon |\psi(x') - \psi(x'')| \le M\omega_{1\nu}, \forall x', x'' \in [a, b]\}, \tag{15}$$

$$\|K_\nu\| \stackrel{\mathrm{df}}{=} \max_{a \le x, t \le b} |K_\nu(x, t)|, \qquad g_\nu = \frac{1}{2} \sum_{j=1}^{m_\nu} |p_j^{(\nu)}|. \tag{16}$$

Then the operator equation (7^1) has a unique solution, and it is a polynomial of the form (10) whose coefficients are found from the system (11) by the method of successive approximations.

Furthermore, the iterative process for finding the coefficients $\{c_k\}_0^n$ converges to the solution of (11) at the rate of a geometric progression with ratio q.

PROOF. Starting from the operator equation (7^1), we define the operator $\mathcal{A}$ on the class $C_{[a,b]}$ of continuous function on $[a, b]$ by the formula

$$\mathcal{A}u = \mathcal{U}_n(\{S_{m_1 i}[K_1 F(u)] + S_{m_2 i}[K_2 F(u)]\}_{i=1}^{N}; x) \quad \forall u \in C_{[a,b]}.$$

Let us see that $\mathcal{A}$ is a contraction operator on $C_{[a,b]}$. Indeed, let u_1 and u_2 be arbitrary functions in $C_{[a,b]}$. Then, by (8)–(10),

$$\|\mathcal{A}u_1 - \mathcal{A}u_2\| = \|\mathcal{U}_n(\{S_{m_1 i}[K_1 F(u_1) - K_1 F(u_2)]$$

$$+ S_{m_2 i}[K_2 F(u_1) - K_2 F(u_2)]\}_{i=1}^{N}; x)\|$$

$$= \left\| \mathcal{U}_n\left(\left\{\frac{x_i - a}{2} \sum_{j=1}^{m_1} p_j^{(1)} K_1(x_i, t_{ij}^{(1)})[F(t_{ij}^{(1)}; u_1) - F(t_{ij}^{(1)}; u_2)] \right.\right.\right.$$

$$\left.\left.\left. + \frac{b - x_i}{2} \sum_{j=1}^{m_2} p_j^{(2)} K_2(x_i, t_{ij}^{(2)})[F(t_{ij}^{(2)}; u_1) - F(t_{ij}^{(2)}; u_2)] \right\}_{i=1}^{N}; x\right)\right\|. \tag{17}$$

In (17) let

$$\mathcal{F}_\nu(x) \stackrel{\mathrm{df}}{=} \frac{d^{(\nu)} - x}{2}(-1)^\nu \sum_{j=1}^{m_\nu} p_j^{(\nu)} K_\nu(x, t_{ij}^{(\nu)})[F(t_{ij}^{(\nu)}; u_1) - F(t_{ij}^{(\nu)}; u_2)], \qquad (18)$$

where $\nu = 1, 2$, $d^{(1)} \stackrel{\mathrm{df}}{=} a$, and $d^{(2)} = b$.

We estimate the modulus of continuity of the functions $\mathcal{F}_\nu(x)$.

If, for example, $\nu = 1$, then (14), (16), and (13) give us that

$$|\mathcal{F}_1(x + \delta) - \mathcal{F}_1(x)| = \left| \sum_{j=1}^{m_1} p_j^{(1)} \left[\frac{x + \delta - a}{2} K_1(x + \delta, t_{ij}^{(1)}) - \frac{x - a}{2} K_1(x, t_{ij}^{(1)}) \right] \right.$$

$$\left. \times [F(t_{ij}^{(1)}; u_1) - F(t_{ij}^{(1)}; u_2)] \right|$$

$$\leq \frac{1}{2} \sum_{j=1}^{m_1} |p_j^{(1)}| \{(x - a)[K_1(x + \delta, t_{ij}^{(1)}) - K_1(x, t_{ij}^{(1)})] + \delta|K_1(x + \delta, t_{ij}^{(1)})|\}$$

$$\times [F(t_{ij}^{(1)}; u_1) - F(t_{ij}^{(1)}; u_2)]$$

$$\leq A g_1 \|u_1 - u_2\|[\omega_1(K_1, \delta)(b - a) + \delta\|K_1\|] \leq A_1 \omega_{11}(K_1, \delta),$$

where $A_1 \stackrel{\mathrm{df}}{=} A g_1 \|u_1 - u_2\|$.

Thus, $\mathcal{F}_1(x) \in A_1 H^{\omega_{11}}$ by (15), and, similarly, $\mathcal{F}_2(x) \in A_2 H^{\omega_{12}}$, where $A_2 \stackrel{\mathrm{df}}{=} A g_2 \|u_1 - u_2\|$. In view of this, we find that

$$|\mathcal{A}u_1 - \mathcal{A}u_2| \leq |\mathcal{U}_n(\mathcal{F}_1; x) - \mathcal{F}_1(x)| + |\mathcal{U}_n(\mathcal{F}_2; x) - \mathcal{F}_2(x)| + |\mathcal{F}_1(x)| + |\mathcal{F}_2(x)|$$

$$\leq A\left[g_1 \varepsilon(\mathcal{U}_n, H^{\omega_{11}}) + g_2 \varepsilon(\mathcal{U}_n, H^{\omega_{12}})\right.$$

$$\left. + (b - a)(g_1\|K_1\| + g_2\|K_2\|)\right] \|u_1 - u_2\|$$

$$\leq q\|u_1 - u_2\| \qquad (19)$$

by considering the linearity of the operator $\mathcal{U}_n$ and by (17), (18), (6), and (12).

This gives us the theorem, by the contraction mapping principle (see, for example, [11], §1.1).

We remark that Theorem 1 in [8] is a consequence of this theorem for $K_1(x, t) = 1$ and $K_2(x, t) = 0$.

§2. Estimates of the deviation

THEOREM 2. *Under the conditions of Theorem 1 the polynomial $y_n(x)$ which solves equation (7^1) approximates the solution of (1) in such a way that*

$$\|y(x) - y_n(x)\| \leq (1 - q)^{-1}[\|y(x) - \mathcal{U}_n(y; x)\| + R_{m_1 n} + R_{m_2 n}], \qquad (20)$$

in which

$$R_{m_\nu n} \stackrel{\mathrm{df}}{=} \|\mathcal{U}_n\| \max_{1 \leq i \leq N} |R_{m_\nu i}[K_\nu F(y)]|, \qquad (21)$$

where the $R_{m_\nu i}[\cdot]$ are the remainders for the quadrature formulas (8) used, while q and $K_\nu F(y)$ are defined by (12) and (9), respectively, and $\|\mathcal{U}_n\|$ is the norm of $\mathcal{U}_n$ in the uniform metric.

PROOF. Starting from (1), (2), (4), (8), and (9), we represent the polynomial in the form

$$\mathcal{U}_n(y;x) = \mathcal{U}_n(\{f(x_i) + S_{m_1 i}[K_1 F(y)] + S_{m_2 i}[K_2 F(y)]$$
$$+ R_{m_1 i}[K_1 F(y)] + R_{m_2 i}[K_2 F(y)]\}_{i=1}^N; x). \qquad (22)$$

According to (22) and (17), termwise subtraction of (7^1) from (1) gives us that

$$|y(x) - y_n(x)| \le |y(x) - \mathcal{U}_n(y;x)| + |\mathcal{U}_n(y;x) - \mathcal{U}_n(f;x)$$
$$- \mathcal{U}_n(\{S_{m_1 i}[K_1 F(y_n)] + S_{m_2 i}[K_2 F(y_n)]\}_{i=1}^N; x)|$$
$$= |y(x) - \mathcal{U}_n(y;x)|$$
$$+ |\mathcal{U}_n(\{S_{m_1 i}[K_1 F(y) - K_1 F(y_n)] + S_{m_2 i}[K_2(F(y) - F(y_n))]\}_{i=1}^N; x)|$$
$$+ |\mathcal{U}_n(\{R_{m_1 i}[K_1 F(y)] + R_{m_2 i}[K_2 F(y)]\}_{i=1}^N; x)|$$
$$= |y(x) - \mathcal{U}_n(y;x)| + |\mathcal{A}y - \mathcal{A}y_n|$$
$$+ |\mathcal{U}_n(\{R_{m_1 i}[K_1 F(y)] + R_{m_2 i}[K_2 F(y)]\}_{i=1}^N; x)|.$$

Estimating the term $|\mathcal{A}y - \mathcal{A}y_n|$ with the help of (19), we find by (21) that

$$|y(x) - y_n(x)| \le \|y(x) - \mathcal{U}_n(y;x)\| + q\|y(x) - y_n(x)\| + R_{m_1 n} + R_{m_2 n},$$

which yields inequality (20). The theorem is proved.

THEOREM 3. *Suppose that in (1) conditions 1)–3) are satisfied for the functions f, F, and K, and the series*

$$1 + A\left\|\int_a^b |K(x,t)|\,dt\right\| + A^2\left\|\int_a^b |K(x,t)|\,dt \int_a^b |K(t,t_1)|\,dt_1\right\| + \cdots$$
$$+ A^k\left\|\int_a^b |K(x,t)|\,dt \int_a^b |K(t,t_1)|\,dt_1 \cdots \int_a^b |K(t_{k-2},t_{k-1})|\,dt_{k-1}\right\| + \cdots$$

converges to some number B [5]. Then for any summator operator $\mathcal{U}_n$ satisfying condition (5) and any quadrature formulas of the form (8) which are exact for mth-degree polynomials and have the property that the operator equation (7^1) has a unique solution $y_n(x)$ $\forall n = 1, 2, \ldots$ such that $\gamma_n \overset{\mathrm{df}}{=} (b - a)AB\varepsilon(\mathcal{U}_n, H_k^\omega) < 1$, we have the estimate

$$\|y(x) - y_n(x)\| \le (1 + \alpha_n)B[\|y(x) - \mathcal{U}_n(y)\| + \|\mathcal{U}_n\|\sigma_{mn}], \qquad (23)$$

in which

$$\alpha_n \overset{\mathrm{df}}{=} \frac{\gamma_n}{1 - \gamma_n}, \qquad (24)$$

and

$$\sigma_{mn} \overset{\mathrm{df}}{=} c_1 E_m^*[K_1 F(y_n)] + c_2 E_m^*[K_2 F(y_n)], \qquad (25)$$

where

$$c_\nu = (b-a)(1+g_\nu), \qquad \nu = 1,2, \tag{26}$$

the g_ν are defined in (16),

$$E_m^*[K_\nu F(y_n)] \overset{\mathrm{df}}{=} \max_{1 \le i \le N} E_m[K_\nu(x_i,t)F(t,y_n(t))] \tag{27}$$

and $E_m[\psi]$ denotes the best approximation of a function $\psi(x)$ by polynomials of degree at most m on $[a,b]$.

PROOF. According to (4) and (7)–(9), we represent (7^1) in the form

$$y_n(x) = \mathcal{U}_n(f;x) + \mathcal{U}_n\left(\left\{\int_a^{x_i} K_1(x_i,t)F(t,y_n(t))\,dt + \int_{x_i}^b K_2(x_i,t)F(t,y_n(t))\,dt \right.\right.$$
$$\left.\left. + R_{m_1 i}[K_1 F(y_n)] + R_{m_2 i}[K_2 F(y_n)]\right\}_{i=1}^N ; x\right)$$

$$= \mathcal{U}_n(f;x) + \mathcal{U}_n\left(\left\{\int_a^b K(x_i,t)F(t,y_n(t))\,dt\right\}_{i=1}^N ; x\right)$$
$$+ \mathcal{U}_n(\{R_{m_1 i}[K_1 F(y_n)] + R_{m_2 i}[K_2 F(y_n)]\}_{i=1}^N ; x).$$

If we now subtract this equality termwise from (1), we get the following chain of inequalities:

$$|y(x) - y_n(x)| \le |y(x) - \mathcal{U}_n(y;x)|$$
$$+ \left|\mathcal{U}_n\left(\left\{\int_a^b K(x_i,t)[F(t,y(t)) - F(t,y_n(t))]\,dt\right\}_{i=1}^N ; x\right)\right.$$
$$\left. + \mathcal{U}_n(\{R_{m_1 i}[K_1 F(y_n)] + R_{m_2 i}[K_2 F(y_n)]\}_{i=1}^N ; x)\right|$$

$$\le |y(x) - \mathcal{U}_n(y;x)|$$
$$+ \left|\mathcal{U}_n\left(\left\{\int_a^b K(x_i,t)[F(t,y(t)) - F(t,y_n(t))]\,dt\right\}_{i=1}^N ; x\right)\right|$$
$$+ \|\mathcal{U}_n\|\left\{\max_{1 \le i \le N} |R_{m_1 i}[K_1 F(y_n)] + R_{m_2 i}[K_2 F(y_n)]|\right\}. \tag{28}$$

Using the remainder estimate

$$R_{m_\nu i}[K_\nu F(y_n)] \le (1+g_\nu)(b-a)E_m[K_\nu(x_i,t)F(t,y_n(t))]$$

for the quadrature formulas, which are exact for mth-degree polynomials (see [13], §3.3), where g_ν is defined by (16), $E_m[\psi]$ is the best approximation of ψ by polynomials of degree at most m in $C_{[a,b]}$, and using the fact that the function

$$\beta_n(x) \overset{\mathrm{df}}{=} \int_a^b K(x,t)[F(t,y(t)) - F(t,y_n(t))]\,dt \tag{29}$$

satisfies the inequality (see [5])

$$\|\beta_n(x) - \mathcal{U}_n(\beta_n, x)\| \le A(b-a)\|y(x) - y_n(x)\|\varepsilon(\mathcal{U}_n, H_K^\omega)$$

we get by (28), (16), (26), (27), (29), and (25) that

$$
\begin{aligned}
|y(x) - y_n(x)| &\le |y(x) - \mathcal{U}_n(y; x)| \\
&\quad + \|\mathcal{U}_n\| \Big\{ c_1 \max_{1 \le i \le N} E_m[K_1(x_i, t)F(t, y_n(t))] \\
&\qquad\quad + c_2 \max_{1 \le i \le N} E_m[K_2(x_i, t)F(t, y_n(t))] \Big\} \\
&\quad + |\mathcal{U}_n(\beta_n, x) - \beta_n(x)| + |\beta_n(x)| \\
&\le \|y(x) - \mathcal{U}_n(y; x)\| + \|\mathcal{U}_n\|\sigma_{mn} + A(b-a)\varepsilon(\mathcal{U}_n, H_K^\omega)\|y(x) - y_n(x)\| \\
&\quad + A\int_a^b |K(x,t)|\,|y(t) - y_n(t)| \\
&\overset{\mathrm{df}}{=} a_n + A\int_a^b |K(x,t)|\,|y(t) - y_n(t)|\,dt. \qquad (30)
\end{aligned}
$$

From this we get step by step that

$$|y(x) - y_n(x)| \le [\|y(x) - \mathcal{U}_n(y; x)\| + \sigma_{mn}\|\mathcal{U}_n\|]B/(1 - \gamma_n)$$

in a way analogous to that in [5] by substituting the right-hand side of (3) for $|y(t) - y_n(t)|$ in the integand.

This implies inequality (23).

The theorem is proved.

We note that estimates of $\|y(x) - \mathcal{U}_n(y; x)\|$ on the right-hand side of (16) have been thoroughly investigated in the theory of approximation of functions for the most familiar summator operators (see, for example, [12]–[16]).

§3. A computation algorithm

The scheme presented in §2 (and in [17]) is realized as a computer program when the operators $\mathcal{U}_n$ are taken to be interpolation operators and the quadrature formulas to be formulas with $m_1 = m_2 = n+1$ nodes at the extremal points of the Tchebycheff polynomials of the first kind:

$$x_i = (a+b)/2 + 0.5(b-a)\cos i\pi/n, \qquad i = 0, \ldots, n. \qquad (31)$$

Such operators and quadrature formulas enable us to obtain a solution in the form of an algebraic polynomial

$$y_n(x) = \sum_{k=0}^{n} c_k(x - x_c)^k, \qquad x_c = \frac{a+b}{2}. \qquad (32)$$

The estimate

$$\|\mathcal{U}_n\| \le \frac{2}{\pi}\ln n + 1 \qquad \forall n = 1, 2, \ldots \qquad (33)$$

is valid in the uniform metric for the norm of the operator $\mathcal{U}_n$ with nodes of the form (31), ([18], p. 18). By the Lebesgue inequality for the interpolation operators and by the direct theorem of Jackson for algebraic polynomials on the class H^ω (see, for example, [15], p. 82), the quantities $\varepsilon(\mathcal{U}_n, H_K^\omega)$ and $\varepsilon(\mathcal{U}_n, H^{\omega_1\nu})$ satisfy the inequalities

$$\varepsilon(\mathcal{U}_n, H_K^\omega) \le 2\omega_{CL}\left(K, \frac{\pi(b-a)}{2(n+1)}\right)\left(\frac{\ln n}{\pi} + 1\right), \tag{34}$$

$$\varepsilon(\mathcal{U}_n, H^{\omega_1\nu}) \le 2\omega_{1\nu}\left(K, \frac{\pi(b-a)}{2(n+1)}\right)\left(\frac{\ln n}{\pi} + 1\right) \tag{34^1}$$

THEOREM 4. *If $y(x) \in MW^\mu \overset{\mathrm{df}}{=} \{\psi\colon \psi C^\mu \wedge \|\psi^{(\mu)}(x)\|_C \le M\}$ and $\|\partial^\mu K(x,t)F(t,y_n(t))/\partial t^\mu\| \le M$, then under the conditions of Theorem 3 the estimate*

$$\|y(x) - y_n(x)\| \le (1 + \alpha_n)A_\mu B\left(\frac{b-a}{2n}\right)^\mu\left\{\left[\left(\frac{2}{\pi}\ln n + 1\right)(1 + 2(b-a))\right] + 1\right\}$$

is valid when interpolation operators and quadrature formulas with the nodes (31) are used, where A_μ is the constant in the Jackson theorem, $n = \mu, \mu+1, \ldots,$ and

$$\alpha_n \le \frac{r_n}{i - r_n}, \quad r_n \overset{\mathrm{df}}{=} 2(b-a)AB\omega_{CL}\left(K, \frac{(b-a)\pi}{2(n+1)}\right)\left(\frac{1}{\pi}\ln n + 1\right). \tag{35}$$

The theorem follows from Theorem 3, Lebesgue's inequality, Jackson's theorem for functions in the class MW^μ, and (33) and (34).

We note that in the case when $\omega_{CL}(K, \delta) = o(1/\ln(1/\delta)) \; \forall \delta \in [0, b-a]$, i.e., when H_K^ω coincides with the class of functions satisfying a Dini-Lipschitz condition ([14], p. 156), the quantity α_n tends to zero as $n \to \infty$ in view of (35).

If $y(x) \in MW^{(n+1)}$ and

$$\|\partial^{n+1}K_\nu(x_i,t)F(t,y(t))/\partial t^{n+1}\| \le M, \quad \nu = 1, 2,$$

then on the basis of Theorem 2, estimates for the deviation of the interpolation polynomial [7] and for the remainders of the quadrature formula [19] with nodes of the form (31), and by (12)–(14), (33), and (34^1), and the fact that the weights in the quadrature formula with nodes of the form (31) are positive, it is easy to see that

$$\|y(x) - y_n(x)\| \le (1-q)^{-1}\left(\frac{b-a}{4}\right)^{n+1} \cdot \frac{4M}{(n+1)!}[1 + 2(2\ln n + \pi)], \tag{35}$$

where

$$q = 2A\left(\frac{\ln n}{\pi} + 1\right)\left[\omega_1\left(K_1, \frac{(b-a)\pi}{2(n+1)}\right) + \omega_1\left(K_2, \frac{(b-a)\pi}{2(n+1)}\right)\right]$$
$$+ (b-a)(\|K_1\| + \|K_2\|). \tag{36}$$

In investigating the complexity of realization for numerical methods it is common (see, for example, [7]) to take the computation of the value of a known function at a single point as an operation. It is obviously necessary to perform $(n+1) + 2(n+1)^2$ such operations at each step of the iteration process in the solution of the integral equation (1)–(2) by the algorithm under consideration.

Following the scheme in [7], we assume that the functions $f(x)$, $K_\nu(x,t)$, and $F(t,y(t))$ in (1) and (2) are such that the solution $y(x)$ and the functions $K_\nu(x,t)F(t,y(t))$ (with respect to the variable t) admit analytic extensions from $[a,b]$ to some domain G bounded by a closed Jordan curve Γ whose distance to $[a,b]$ is equal to R, with the extensions bounded on Γ by the number M.

Such functions $\psi(x)$ satisfy (see [7]) the inequality

$$\|\psi^{(n+1)}(x)\| \le (n+1)!M/R^{n+1}.$$

Replacing M by $(n+1)!M/R^{n+1}$ in (35), we conclude that for the error of the polynomial approximation in this case not to exceed a given precision ε it is necessary to take the number of nodes in the algorithm to be such that

$$(1-q)^{-1}\left(\frac{b-a}{4R}\right)^{n+1}4M(1+2(2\ln n + \pi)) \le \varepsilon/2, \tag{37}$$

i.e.,

$$n = \left[\left(\ln \frac{\varepsilon(1-q)}{8(1+4\ln n + 2\pi)M}\right) \Big/ \left(\ln \frac{b-a}{4R}\right)\right] \stackrel{\mathrm{df}}{=} N_1(\varepsilon,n). \tag{38}$$

The inequality

$$\|y_n^k(x) - y_n(x)\| < \varepsilon/2$$

is satisfied at the kth step of the iteration process for the chosen initial approximation $y_n^0(x)$ [7] in view of the relation

$$\|y_n^k(x) - y_n(x)\| \le q^k\|y_n^0(x) - y_n(x)\| \stackrel{\mathrm{df}}{=} q^k\varepsilon_0,$$

if

$$k = [\ln(\varepsilon/2\varepsilon_0)/\ln q] + 1. \tag{39}$$

The next theorem can be seen from (37)–(39).

THEOREM 5. *For the polynomials constructed according to the above algorithm to approximate the solution of* (1) *and* (2) *with a given accuracy ε it suffices that the number N of operations be equal to*

$$N[(n+1) + 2(n+1)^2]k = [N_1(\varepsilon,n) + 1][2N_1(\varepsilon,n) + 3][\ln(\varepsilon/2\varepsilon_0)/\ln q] + 1),$$

i.e., $N \asymp |\ln^3 \varepsilon|$ to within the principal terms.

We remark that the simplicity with which the coefficients in the interpolation polynomial [8] and the weights in the quadrature formula (see [19]) are computed enables us to realize this algorithm even on the small computers in the MIR series.

§4. Numerical examples

Let us consider some of the model examples which have served for controls and checks on the efficiency of computer programs and for comparison of our algorithms with other algorithms.

The computations were performed on a MIR-2 computer to 10 decimal places.

In the Ex. 1 and Ex. 2 rows of Table 1 (for Examples 1 and 2, respectively) we give the maximal errors ε_n of the polynomial approximation for $m_1 = m_2 = n + 1 = 3, \ldots, 9$ at the points $x_i = a + (b - a)i/20$, $i = 0, \ldots, 20$, i.e., the quantities

$$\varepsilon_n = \max_i |y(x_i) - y_n(x_i)|. \tag{40}$$

n $\diagdown$ $\mathcal{N}_n/n$	2	3	4	5	6	7	8
Ex. 1	$1.2 \cdot 10^{-1}$	$4 \cdot 10^{-2}$	$3 \cdot 10^{-3}$	$6.4 \cdot 10^{-4}$	$4.8 \cdot 10^{-5}$	$2.2 \cdot 10^{-6}$	$3.7 \cdot 10^{-7}$
Ex. 2	$8.1 \cdot 10^{-1}$	$2 \cdot 10^{-1}$	$6 \cdot 10^{-2}$	$7.8 \cdot 10^{-3}$	$2.1 \cdot 10^{-5}$	$1 \cdot 10^{-8}$	$2.1 \cdot 10^{-9}$

$$\text{TABLE 1}$$

EXAMPLE 1. The Hammerstein integral equation

$$y(x) = (x - 3)/2 - 2 \int_1^2 K(x,t)\,dt/(t^4 y(t)), \tag{41}$$

is solved, with

$$K(x,t) = \begin{cases} (2 - x)(t - 1) & \text{for } 1 \leq t \leq x \leq 2, \\ (2 - t)(x - 1) & \text{for } 1 \leq x \leq t \leq 2; \end{cases}$$

the exact solution is $y(x) = -1/x$.

As a result of the computation when, for example, $m_1 = m_2 = n+1 = 8$ in (11) and with specified accuracy $\varepsilon = 10^{-8}$, the following array $\{c_k\}_0^7$ of coefficients of the polynomial $y_7(x)$ of form (32) was output on the printout:

$$\{c_k\}_0^7 = \{-0.6666647095, 0.4444431984, -0.296501099, 0.1976671084,$$
$$-0.1283479095, 0.0855701577, -0.0760507245, 0.0506809783\}.$$

EXAMPLE 2. The following example, which well illustrates the advantages of the proposed algorithm for solving (1)–(2) over the algorithm worked out in [4], was taken from Atkinson's paper [4]:

$$y(x) = f(x) - \lambda \int_a^b K(x,t)y(t)\,dt, \tag{42}$$

where

$$K(x,t) = \begin{cases} (x - 1)t & \text{for } 0 \leq t \leq x \leq 1, \\ (t - 1)x & \text{for } 0 \leq x \leq t \leq 1, \end{cases}$$

and the free term $f(x)$ is such that the function $y(x) = \mu^2(1 - x)x^\mu$ is a solution of (42).

The errors in the solution of this equation by the algorithm in [4] (Table X) with 256 nodes in the Gauss quadrature formula and with $\lambda = -10$ and $\lambda = -30$ were equal to $1.4 \cdot 10^{-3}$ and $2.1 \cdot 10^{-4}$, respectively.

We remark that if the proposed algorithm is applied to (42), then we need to take a quadrature formula and an interpolation operator with 8 nodes to obtain an accuracy of 10^{-8}.

The coefficients in this polynomial have the form

$$\{c_k\}_0^7 = \{0.3906250019, 3.125000008, 7.812499756, -0.0000005411,$$
$$- 31.24999713, -49.99999372, -25.000000790, -0.0000168657\}.$$

EXAMPLE 3. Consider the equation

$$y(x) = e^{-x} + \int_0^{0.1} K(x,t)y^2(t)\,dt,$$

where $K_1(x,t) = e^{t-x}$ and $K_2(x,t) = 0$; this was solved in [1] (p. 31) like a Volterra equation. Its exact solution is $y(x) = 1$.

When, for example, $n = 2$ and the specified accuracy was $\varepsilon = 10^{-5}$, the result of the computation was the second-degree algebraic polynomial

$$y_2(x) = 1.00000146 + 0.00095720(x - 0.05) + 0.00132860(x - 0.05)^2,$$

whose maximal deviation from the exact solution at the points $x_i = 0.05i$ ($i = 0,\ldots,10$) is $\varepsilon_2 = 1.59 \cdot 10^{-6}$. Comparing this error with that in Table 6 of [1] (p. 32) (the maximal error is $1.6 \cdot 10^{-2}$), we conclude that our algorithm gives sufficiently high accuracy even in the case of kernels discontinuous on the line $x = t$.

REMARK 1. The behavior of the quantities $\{\varepsilon_n\}_2^\infty$ in Table 1 with increasing n shows the high efficiency of our algorithm, as well as the good agreement of the theoretical predictions in Theorem 4 and inequality (35) with the results of the numerical experiments.

REMARK 2. It was mentioned in [1] (p. 28) that application of the method of mechanical quadratures to the solution of integral equations with variable upper limit involves a number of special features and that the trapezoid formula should be expected to be preferable here. This remark obviously applies also to equations (1)–(2) and, consequently, for methods presupposing computation of the integral in such equations on the whole interval $[a, b]$ (for the method of mechanical quadratures, the Nyström-Atkinson method, and others; see [1]–[4]) it is impossible in this case to guarantee accuracy higher than $O(h^2)$, where h is the step in the trapezoid quadrature formula.

This demonstrates the expediency of solving equations of the form (1)–(2) by the scheme in §1, of which the projection methods with perturbation [11] make up a special case.

EXAMPLE 4. To check the efficiency of the proposed algorithm in the solution of practical problems let us consider one of the basic equations of atmospheric optics (see [3] and [20]):

$$S(t) = f(t) + \frac{\lambda}{2} \int_0^h E(|t - \tau|)s(\tau)\,d\tau, \tag{43}$$

where the kernel is the integral exponential

$$E(\mu) = \int_0^1 e^{-\mu/z}\,dz/z, \qquad \mu \geq 0.$$

This equation simulates the process of radiative transfer in a plane-parallel homogeneous medium of optical thickness h and with probability λ that a quantum of radiation survives collision with a particle of the medium, where the function $f(t)$ is some external action.

The equation can be solved on a computer when the parameters have the following values: $\lambda = 0.5$, $h = 0.3$, and $f(t) = \exp[2(t - 0.3)]/8$ (see [20], Chapter IV, §12).

Table 2 gives computational results obtained by Sobolev and Minin (SM) and by the invariant imbedding method (IIM) of [20] (loc. cit.), along with the values of the polynomials of the form (21) obtained by our proposed algorithm, in which the function $E(\mu)$ is approximated according to the recommendations in [3] (§7) by means of the Gauss quadrature formula on $[0, 1]$ of order six with $n = 3, 4, 5, 6$.

t	SM	IIM	Proposed algorithm			
			$n = 3$	$n = 4$	$n = 5$	$n = 6$
0.0	0.083	0.082559	0.082393	0.082519	0.82525	0.082533
0.1	0.103	0.102958	0.103135	0.103039	0.102966	0.102987
0.2	0.123	0.122851	0.123064	0.123070	0.122859	0.122899
0.3	0.141	0.140966	0.140709	0.140939	0.140935	0.140950

TABLE 2

Analyzing the computational results in this table, we see that our algorithm enables us to obtain sufficiently high accuracy with significantly smaller volume of computational work than in the invariant imbedding method.

For example, for $n = 4$ the polynomial

$$S_4(t) = 0.1128690675 + 0.2011567277(t - 0.15) + 0.0551584839(t - 0.15)^2$$
$$+ 0.2903161637(t - 0.15)^3 - 4.739997267(t - 0.15)^4$$

obtained by the algorithm in §3 deviates by no more than $\varepsilon = 0.00022$ from the solution of the invariant imbedding method in which the fourth-order Adams-Moulton procedure was used with integration step $\Delta = 0.005$.

We remark that, first and foremost, the fact that our algorithm enables us to obtain a solution in a simple analytic form should be included in its advantages over the invariant imbedding method. This is especially important for the radiative transfer problem in connection with the fact that the solution is used for subsequent computations in the final determination of the main desired function in this problem: the radiation intensity [3].

Moreover, the program and the working arrays, as well as the array for storing the coefficients of the polynomial, require a considerably smaller memory capacity than the program for the invariant imbedding method ([20], Appendix A).

The author thanks V. K. Dzyadyk for posing the problem and for his attention to this work, and P. N. Denisenko for his attention to the computation aspects of the work.